The Computer Modelling of
Mathematical Reasoning

The Computer Modelling of Mathematical Reasoning

Alan Bundy

*Department of Artificial Intelligence,
University of Edinburgh*

1983

ACADEMIC PRESS, INC.

(Harcourt Brace Jovanovich, Publishers)

London   Orlando   San Diego   New York
Toronto   Montreal   Sydney   Tokyo

ACADEMIC PRESS INC. (LONDON) LTD.
24/28 Oval Road
London NW1

United States Edition published by
ACADEMIC PRESS, INC.
Orlando, Florida 32887

British Library Cataloguing in Publication Data

Bundy, Alan
  The computer modelling of mathematical reasoning.
  1. Artificial intelligence
  I. Title
  001.53′5   Q335

  ISBN 0-12-141250-4 (Cased)
  ISBN 0-12-141352-0 (Paperback)

PRINTED IN THE UNITED STATES OF AMERICA

85 86 87 88   9 8 7 6 5 4 3 2

Preface

This book started as notes for a postgraduate course in Mathematical Reasoning given in the Department of Artificial Intelligence at Edinburgh from 1979 onwards. Students on the course are drawn from a wide range of backgrounds: Psychology, Computer Science, Mathematics, Education, etc. The first draft of the notes was written during a term's sabbatical leave in 1980. Later they were used for a similar course at undergraduate level.

While there are now several textbooks on Artificial Intelligence techniques and, more particularly, on problem solving and theorem proving, I felt the need for a book concentrating on applications of these techniques to mathematics. There was certainly enough material, but it was scattered in research journals, conference proceedings and theses. If it were collected together I hoped it might prove of interest to a wider audience than the usual artificial intelligentsia; I hoped that mathematicians and educationalists might find it a eye opener to how computational ideas could shed light on the process of *doing* Mathematics.

I also hoped to give more unity to some, rather disparate, pieces of research. In particularly, I wanted to show how the, so called, 'non-Resolution theorem proving' techniques could be readily brought into a Resolution framework, and how this helped us to relate the various techniques — creating coherence from confusion. In order to achieve this goal I have taken strong historical liberties in my descriptions of the work of Boyer and Moore, Gelernter, Lenat, etc. I have redescribed their work in a uniform framework, ignoring aspects of peripheral interest, and focusing on what I take to be their essential contribution. I call such descriptions, *rational reconstructions*. This does not imply that the original work was irrational — only that my reconstructions are rational. I apologise to any of the rationally reconstructed who feel mistreated. My excuse is that the reworking of research into a coherent whole is a vital, but neglected, part of Artificial Intelligence research, and that it is better to have tried and failed than never to have tried at all.

Reading Strategies

It is not necessary to be a professional mathematician or computer scientist to read this book, but the book does presuppose some mathematical knowledge. For instance, it is necessary to know what a set and a group are.

v

I have endeavoured to make it fairly self contained, e.g. by including an introduction to mathematical logic. But self-containedness brings its own problems; if the book is not to be too long, then the pace must not be too slow. I have tried to get the reader quickly to the heart of the book — the techniques of automatic inference — without losing him on the way.

– Chapter 1 is a general introduction; it motivates the subject and gives some of the historical background.

– Part I is a three chapter introduction to mathematical logic; it describes only those aspects of logic that are required to understand the rest of the book, and may be omitted by anyone who understands elementary predicate logic.

– Part II is a three chapter introduction to Resolution theorem proving. It may be omitted by anyone who knows what SL Resolution is.

– Part III consists of five rational reconstructions of theorem proving techniques or programs. Each was selected because it contributes an important partial solution to the problem of guiding the search for a proof. This part is the heart of the book.

– Part IV is a two chapter discussion of aspects of mathematical reasoning other than proving theorems — although they both reduce to theorem proving in the end.

– Part V is a three chapter investigation of the more mathematical aspects of theorem proving, e.g. completeness proofs. It may have to be omitted by those without a good background in mathematics.

– The last chapter discusses some applications of the techniques described in the book, from algebraic manipulation to education.

– The appendices include: computer programs, notational discussions and solutions to exercises.

Scattered throughout the book are exercises of an elementary nature. Readers may want to use these to test their understanding of the text. Some of the exercises contain material that is drawn on later in the book. Solutions may be found in appendix IV.

Readers already familiar with the literature of mathematical reasoning may be particularly interested in the nonstandard presentations of

Gelernter's Geometry Machine, in chapter 10, and Lenat's AM, in chapter 13. Section 10.5 onwards, of chapter 10, contains new results.

Acknowledgements

I would like to thank: Gordon Plotkin, who was an untiring source of information; Richard O'Keefe, Leon Sterling, Alan Borning, David Plummer, Roy Dyckhoff, Lincoln Wallen, Henry Thompson and Alberto Peterossi, who helped me debug the drafts; and Bob Boyer, J Moore, Gerard Huet, Woody Bledsoe, Robert Shostak and Dough Lenat, who kindly read the chapters describing their work and gave me invaluable feedback. The students of the Department of Artifical Intelligence at Edinburgh were involuntary guinea pigs. The Edinburgh DEC10, the FINE editor and the SCRIBE text formatter also assisted.

May 1983   Alan Bundy

Dedication

This book is dedicated to Bernard Meltzer who inspired me to build artificial mathematicians and created the ideal atmosphere for me to do so.

Contents

III. Guiding Search

V. Technical Issues

1
Introduction

☐ Sections 1.1 and 1.2 provide motivation for studying mathematical reasoning.

☐ Section 1.3 gives the background of the field, describing its origins in Logic (section 1.3.1) and Psychology (section 1.3.2), and giving a short history of the automation of inference (section 1.3.3.)

1.1 Why Read This Book?

This book is aimed at people interested in the question

How do you do Mathematics?

i.e. at professional mathematicians, students of Mathematics, teachers of Mathematics, psychologists studying mathematical reasoning and anyone else who is curious about the apparently mysterious processes by which mathematicians: conjecture theorems, formulate definitions, construct proofs and build mathematical models. My theme is that light can be shed on these 'mysterious' processes with the aid of a wonderful tool – the digital computer. By building computer programs which 'do' mathematics we can explore how it is possible to do mathematics; what the vital talents are that separate success from failure and how we can all learn to be better mathematicians.

The building of computer programs for doing mathematics is part of the new science of Artificial Intelligence. The aim of Artificial Intelligence is to study all aspects of intelligence by 'computational modelling', mathematical reasoning being just one such aspect of intelligence. Other aspects which are studied include: the ability to coordinate hand and eye to manipulate objects; the ability to hold a conversation in a so-called 'natural' language like English (as opposed to an artificial programming language like ALGOL or FORTRAN) or the ability to diagnose an illness and prescribe a cure.

Whatever aspect of intelligence you attempt to model in a computer program – the stacking of bricks, a cocktail party conversation or the proving of the compactness theorem – the same needs arise over and over again.

- The need to have knowledge about the domain.
- The need to reason with that knowledge.
- The need for knowledge about how to direct or guide that reasoning.

1

Mathematical reasoning is a particularly convenient domain for studying intelligence, because the knowledge required is neatly circumscribed and the goals clear and unambiguous.

1.2 What good is Automatic Mathematical Reasoning

If we are successful in building artificial mathematicians, computer programs which do mathematics, what practical benefits will this bring?

The professional mathematician and scientist can expect a range of 'mathematical aids': programs which help him with the more tedious or complex parts of his work, programs which check his proofs, programs which offer suggestions about what to do next. Simple versions of such programs already exist. Many scientists have access to algebraic manipulation systems which can integrate symbolic expressions, simplify formulae and solve equations. We will have more to say about such programs in chapters 12 and 18.

The teacher and student of Mathematics can expect a range of computational theories which will lay bare the more mysterious aspects of the mathematician's art: how a good proof step is chosen from among the possible ones, how interesting conjectures are made, how a mathematical model is made. The teacher may also be interested in models of poor students which explain what it is that a student is doing wrong every time he gets a wrong answer. Such 'diagnostic models' will enable the teacher to design remedial instruction, tailor made, to put the student on the right road again. Again, simple versions of such theories already exist. Some primary school teachers have been given access to a program which simulates a wide range of commonly occurring, faulty, subtraction procedures. The teachers gain experience in diagnosing the fault and learn in few minutes what might otherwise take years: that most arithmetic errors are not due to 'carelessness', but are the result of carefully following a, slightly faulty, arithmetic procedure. We will say more about such programs in chapter 18.

1.3 The Historical Perspective

Of course, workers in Artificial Intelligence are not the first people to have asked how Mathematics is done. Mathematicians themselves have devoted a lot of thought to just this question. Before we start to get involved in the world of computer programs it will behove us to look at the answers they provide, see how adequate they are and how useful for our purposes.

This precomputational work on mathematical reasoning divides into two camps: the normative work of Mathematical Logic, which asks

What are the legal modes of reasoning?

and tries to build a calculus of reasoning, and the psychological work of people who, looking at themselves and others, ask

How do we actually go about doing mathematics?

We will see that both camps have something to offer us. Mathematical Logic will give a start on how to represent the knowledge of Mathematics and how to reason with it. It will not help us to guide that reasoning process, i.e. to decide what to do when. For that help we will turn to the psychological studies, but we will not find our needs satisfied in the kind of detail we would like. Nobody has yet succeeded in giving a precise recipe for how to be a successful mathematician. Instead we will find some good advice and useful observations and hints. These will furnish a starting point.

Let us start by looking at the work in Logic. Space does not allow more than the most superficial survey. In the succeeding chapters we will go a little deeper, but we will only be studying concepts on a 'need to know basis', so the reader wanting to know more should refer to one of the many excellent introductory books on Mathematical Logic, e.g. [Mendelson 64].

1.3.1 Mathematical Logic

This story, as so many stories in Science, starts in Ancient Greece. Aristotle was the first to try to describe the laws of Logic, i.e. the forms that correct arguments could take. He discovered 19 correct forms, and called them *syllogisms*. Here is an example of a syllogism

All Ancient Greeks were perfect.   (i)
Aristotle was an Ancient Greek.
_____

Therefore, Aristotle was perfect.

The two sentences above the line are called the *hypothesis* and the sentence below the line is the *conclusion*. If the hypothesis is true then the conclusion follows from it. Notice that this syllogism is a correct argument form, regardless of whether the hypothesis is true, e.g. whether all Ancient Greeks were perfect or Aristotle was an Ancient Greek. Logicians do not worry about the *content* of arguments – whether the hypothesis is true – but only with the *form* – what sorts of argument are correct.
The form of (i) above is

All Ps were Q.
X was a P.
_____

Therefore, X was Q.

The scholastics who studied and refined Aristotle's work, in the middle ages, gave names to all the valid syllogisms. The one above is called *Darii*.

In our example, P was 'Ancient Greek', Q was 'perfect' and X was 'Aristotle'. We could substitute anything else for P, Q and X and still have a correct form, e.g. 'syllogism' for P, 'invalid' for 'Q' and 'Darii' for X. We can also change the tense from past to present.

> All syllogisms are invalid.
> Darii is a syllogism.
> _____
> Therefore, Darii is invalid.

All the sentences involved in the 19 syllogisms, whether as hypothesis or conclusion, take one of a small number of fixed forms, like 'All As were B.' or 'Some As are B.', so it is not possible to capture with the syllogisms alone, reasoning which involves sentences in other forms.

Thus Aristotle's list of 19 syllogisms do not exhaust the correct argument forms. Unfortunately, due to Aristotle's prestige, it was believed for many years that they did. The scholastics believed in the content, rather than just the form, of syllogism (i) above. This misplaced faith held up the development of logic for about a millennium and it took a considerable effort of will by Boole in the 19th century to add to the accepted correct forms. Here is one of Boole's additions.

> Either all reasoning is syllogistic or Aristotle was wrong.   (ii)
> All reasoning is not syllogistic.
> _____
> Therefore, Aristotle was wrong.

Boole invented Propositional Logic – a mathematical theory covering the way in which elementary sentences (or *propositions*) can be combined with connecting words (or *connectives*) like *and, or, not, if,* etc.

Notice how the first hypothesis of the form (ii) above consists of two propositions: 'all reasoning is syllogistic' and 'Aristotle was wrong', connected together with the connective, 'Either ... or ...'. In fact the form of (ii) is:

> Either *P* or *Q*.
> Not *P.**
> _____
> Therefore, Q.

Boole's Propositional Logic and Aristotle's syllogisms described two disjoint families of correct argument forms, but they still did not exhaust all the correct forms. Both families are too inflexible, but in different ways. The syllogisms only allow reasoning between sentences of a few simple forms.

_____

*In English we often embed the connective 'not' inside the proposition.

We cannot connect together several sentences to form a hypothesis in the way that we can in Propositional Logic. Propositional Logic, on the other hand, can be used to build hypotheses and conclusions of arbitrary complexity from propositions, but we cannot delve inside an proposition itself as we can in a syllogism, e.g. we cannot extract the bits, 'reasoning' and 'syllogistic' from 'All reasoning is syllogistic.'.

The liberation of the correct argument forms was provided by Gottlieb Frege, who invented Predicate Logic. Here the basic building blocks are *objects* and relations (or *predicates*) between them. Predicate Logic combines Propositional Logic's ability to construct new sentences from old, with the syllogistic ability to delve into the internal structure of the sentences. It includes all the correct forms of both its predecessors, and more besides. Here is one of the new forms

> This is an argument.
> This is not propositional.
> This is not syllogistic.
> _____
> Therefore, some arguments are neither propositional nor syllogistic.

Notice how properties of being an argument, a proposition and syllogistic are extracted from inside the three hypothesis sentences and used to construct a new sentence, together with the *quantifier*, some, and the connective, neither.. nor ...

Predicate Logic includes nearly all the correct argument forms we shall have cause to use in this book, although we will occasionally stray into higher order logics. We will be taking it up again in chapters 2 to 4, where we will use it to represent mathematical knowledge and mathematical reasoning. Our justification will be a theorem, due to Jacques Herbrand, which suggests an automatic procedure for finding proofs to theorems, which is guaranteed to find a proof if there is one. The procedure suggested by Herbrand's theorem turns out to be horribly inefficient, but it will serve as a starting point.

1.3.2 Psychological Studies

Mathematical Logic concerns itself with justification rather than discovery. That is, we can use its result to show that an existing mathematical proof is correct, but despite Herbrand's Theorem, it is not a lot of use in actually finding the proof in the first place. Finding a proof (or even deciding what theorem to try and prove) is largely a matter of 'experience', 'luck', 'intuition' and all the other mysterious processes which we sometimes feel are beyond understanding. The purpose of this book is to establish that, on the contrary, these processes can be understood, and even modelled in a computer program. To help establish this, we will now appeal to those few mathematicians who have tried to capture something of their skill in words.

The most famous of these is George Polya. In his book, 'How to Solve It' [Polya 45], he summarized and explained some advice, designed to improve the reader's problem solving ability. This advice took the form of questions for the problem solver to ask himself.

> What is the unknown?
> Do you know a related problem?
> Could you restate the problem?

Generations of students have found these questions an inspiration to their problem solving efforts. They undoubtably help to liberate thought by suggesting new avenues and pointing to the continuation of old ones. How helpful are they to us in building an artificial mathematician?

Unfortunately, Polya's exhortations lack the detail required to make them immediately useful in our task. Consider, for instance, 'Could you restate the problem?'. This question presumes knowledge of English and the way in which problems can be stated in it, which we are only just beginning to give our computers. To be able to make use of a restatement of the problem implies knowing subtle relationships between problem statements and solution methods which our computers just do not know. We must start by building up an armoury of solution methods and relating them to problem types. Then we might be able to build a computer program which could understand Polya's advice – a first step on the road to using it.

All is not lost, however. A later book of Polya's, 'Mathematical Discovery' [Polya 65], holds out more hope. The advice given in this two volume set is much more domain specific. For instance, the first chapter of the first volume gives specific help on making geometric constructions. This advice *has* been embodied in a computer program [Funt 73].

Polya's attitude in trying to understand the 'mysterious' aspects of problem solving is all too rare. The usual attitude of mathematicians is reflected in their published research papers and in mathematics textbooks. Proofs are revamped and polished until all trace of how they were discovered is completely hidden. The reader is left to assume that the proof came to its originator in a blinding flash, since it contains steps which no one could possibly have guessed would succeed. The painstaking process of trial and error, revision and adjustment are all invisible.

The only attempt, of which I am aware, to explain the process by which a proof was constructed, is B.L. van der Waerden's paper, 'How the proof of Baudet's conjecture was found' [Waerden 71]. Here is an excellent description of the process of trial and error, an induction hypothesis is proposed, and gradually modified, as successive attempts to prove the induction step fail. Yet each failure suggests the modifications which take the next attempt a step further.

Imre Lakatos undertook a more ambitious task – to chart the development of a particular theorem through several centuries [Lakatos, 1976]. He describes the trial and error processes which have given us our modern version of Euler's Theorem, that for all polyhedra, $V-E+F = 2$, where V is the number of vertices, E the number of edges and F the number of faces. The conjecture of the theorem from a number of examples, the derivation of a proof, the discovery of a counterexample, the location of the fault in the proof, the revision of the proof and the many forms this revision may take. The theme of his book is the subtle interaction between proof and counterexample: how a flawed proof may suggest a counterexample, how a counterexample may suggest ways to improve a proof. For instance, we may change the definitions of key concepts in the theorem, like the concept of 'polyhedra', in order to exclude the counterexample.

A major implication of Lakatos's discussion, is that the notion of *definition* itself is a subtle one. Not only do different people at different times have different definitions of concepts like polyhedra, but the definition of a single person at a single time may be hazy, so that it cannot be used to decide whether some difficult 'borderline' case is a polyhedron or not.

How useful are the observations of van der Waerden and Lakatos in our task of building an artificial mathematician? Again, their advice, while extremely important, is premature given our current state of development. Workers in Artificial Intelligence have been struggling to build programs to generate sensible first proof attempts, which could form the basis for such refinement processes. Until they succeed at this, the observations of van der Waerden and Lakatos will go unused.

However, we will see in chapters 10 and 11 that there have been some attempts to guide the search for a proof both by the use of counterexamples and by gathering evidence from an earlier failed attempt at a proof, although neither of these computer programs displays the sophistication observed by van der Waerden and Lakatos.

1.3.3 Automatic Theorem Proving

The idea of building an 'artificial mathematician' can be traced back to two sources: the invention of the digital computer and the development of Mathematical Logic. This was compounded by the fact that some very able mathematicians, men like Alan Turing, were engaged in both enterprises. Unfortunately, the psychological observations described above played a very minor role in the early days.

Mathematical Logic provides a formal theory of Mathematics. That is, it shows how any branch of Mathematics can be described as the derivation of theorems from a set of axioms using some rules of inference. These rules of inference are just a selection of the correct argument forms which we illustrated in section 1.3.1 above. The axioms are some sentences which we decide to accept as true and which can then be used as the hypothesis of a rule of inference to derive the conclusion of the rule as a theorem. The classic model for this is Euclidean Geometry.

This suggests a simple procedure for developing a mathematical theory. Starting with just the axioms as the 'theorems' of the theory we can pick a rule of inference at random, find some theorems which can be used as its hypothesis and add its conclusion to our pool of theorems. Thus if we pick as our rule of inference the Darii syllogism

>   All Ps are Q.
>   X is a P.
>   _____
>   Therefore, X is Q.

and we already have in our pool of theorems

>   All odd numbers are prime
>   and
>   9 is an odd number

then we could substitute 'odd number' for P, 'prime' for Q and 9 for X and derive the new theorem

>   9 is prime

Of course, this would be a funny mathematical theory, but we should be able to model correct reasoning from faulty assumptions as well as from correct assumptions.

If we were interested in proving a particular conjecture we could just go on generating new theorems hoping that the one we wanted would turn up. This would be a hopeless method for humans to use. All sorts of irrelevant, and probably uninteresting, theorems would get generated before our conjecture was derived. Even if our conjecture were provable there is no knowing when it might get proved. However, digital computers are known for their speed and their capacity to do oodles of boring work without making mistakes. This procedure might be a practical one for them.

Well it isn't. Despite their impressive speed, the number of possible moves in this mathematical game are too great. Using this technique, a computer can be made to churn out trivial and uninteresting 'theorems' at an enormous rate, but unless the one you are interested in has a particularly simple proof it is unlikely to turn up this side of doomsday.

To see why this is so, consider the 'substitution' rule of inference, which most mathematical theories contain.

$$\frac{A(X)}{A(T)}$$

This is to be read as follows

> If a theorem A contains a variable X then we may substitute any term T for X in A to form a new theorem, A(T).

The catch here is the 'any term T'. If this is a theory about arithmetic then 'any term' might mean any natural number, 0,1,2,3,... etc. and any combination of these with arithmetic operations, $+$, $\times$, $/$ etc. and even other variables. So from the theorem $X \neq X+1$, we may generate $0 \neq 0+1$, $1 \neq 1+1$, $1+2 \neq 1+2+1$, $1+Y \neq 1+Y+1$, and so on. This is an awful lot of new theorems – an infinite number– and worse still those new theorems which contain variables can now be used to generate new ones in their turn.

This is where Herbrand's Theorem comes to the rescue. The effect of his theorem, which is described in more detail in chapter 16, is to limit the 'any terms' that we need consider substituting for X. We can restrict ourselves to terms without variables, and we need only consider those which can be constructed from symbols already occurring in the conjecture and the axioms. It also suggests adding the negation of the conjecture as a new axiom and searching for a contradiction, rather than searching for the conjecture among those generated.

The theorem proving process implied by Herbrand's Theorem was implemented by Paul Gilmore in 1960 [Gilmore 60]. It turned out to be pretty inefficient. It could generate the proofs of a few trivial theorems, but got hopelessly lost when trying to prove anything slightly more complicated.

But Gilmore's program showed that automatic theorem proving was possible in principle. During the rest of the decade there was a concerted attempt to improve the basic Gilmore procedure. The key event in this period was Alan Robinson's invention of the *Resolution* procedure [Robinson 65]. Robinson showed how the standard axioms and rules of inference of Predicate Logic could be replaced with the single, slightly complicated, rule he called resolution. Naturally this rule included elements of the potentially explosive, substitution rule, but in a way made subservient to the context: the only substitutions attempted were those which enabled other rules to be applied. Appropriate substitutions were calculated by a procedure called *unification*. We will study unification and resolution more in parts II and V.

For the remainder of the decade nearly all automatic theorem proving systems were either refinements or extensions of the Resolution procedure. Systems were built with wonderful sounding names, like Hyperresolution,

Paramodulation, P1 Deduction, SL Resolution, Lush Resolution, Connection Graphs. Each of them represented an improvement over the original Resolution procedure and steady progress was made, with ever more difficult theorems being proved. There seemed no reason why this situation could not continue indefinitely, until the whole of Mathematics was gradually conquered by explaining how it can be incorporated into a Resolution framework.

The tranquillity was shattered by strident criticism from other workers in Artificial Intelligence, especially those from the Massachusetts Institute of Technology. They argued that this work, on what they termed 'Uniform Proof Procedures', was not going to conquer the whole of Mathematics, but only a trivially small subset of Mathematics. To generate proofs of interesting, non-trivial theorems required the use of sophisticated, domain specific knowledge, for which no provision was made in the Resolution family of theorem provers.

This attack was at first resisted and then reluctantly accepted. The development of Resolution theorem provers ceased, except in a few isolated pockets of resistance. People began to look at the works of Polya and Lakatos and to introspect about their own mathematical activity in order to get inspiration as to how to proceed. A new family of *Natural Deduction* theorem provers emerged. All sorts of new techniques were attempted: domain specific guidance, the use of modeis and counterexamples, rewrite rules, analogical reasoning. A main goal of this book is to try to explain the work done in this period and to give order to it.

1.4 Summary

This chapter has introduced the building of computer programs to do mathematical reasoning. We have motivated the building of 'artificial mathematicians' and described some of the historical background to the attempt to do so. In subsequent chapters we will consider *how* to do it.

Part I:
Formal Notation

2
Arguments about Propositions

☐ This chapter is an introduction to Propositional Logic.
☐ Section 2.1 introduces the truth functional connectives.
☐ Section 2.2 considers formulae made from connectives and propositions. It defines the meaning of such formulae using semantic trees (section 2.2.1) and shows how to use these to identify correct argument forms (sections 2.2.3 and 2.2.4).

Although those involved in the renaissance of systems and techniques developed in the 1970s were not always aware of their debt, all of them relied heavily on the 1960s work on Resolution. To help give unity to all these later efforts we will recouch them all in the language of Predicate Logic and Resolution. This will involve us in studying sufficient of the field of Mathematical Logic to understand how mathematical knowledge and reasoning can be represented in a mathematical calculus and to understand the significance of Herbrand's Theorem and the resolution rule of inference. This chapter deals with Propositional Logic; the next chapter extends this to Predicate (or First Order) Logic; and the chapter after that extends this again to Omega Order Logic.

Readers familiar with Mathematical Logic may wish to skip now to part II and those familiar with Resolution may wish to skip beyond that to chapter 7. Let me tempt even the experts to stay by announcing that my approach to Logic is a little nonstandard, having been specially designed to support the theoretical demands of automated reasoning.

For those who are still with us we will start our story with logical notation and how it can serve, as a alternative to mathematically flavoured English, as a tool for representing mathematical statements in a form suitable for manipulation by a computer.

2.1 Truth Functional Connectives

In section 1.3.1 we mentioned the Propositional Logic connectives, and, or, not, etc, and how they can be used to connect propositions together. In this section we explore these connectives a little more deeply: defining them properly and illustrating them with examples drawn from Mathematics.

2.1.1 Negation

In Mathematics we have a variety of ways of saying that something is not the case. We may use English and say

n is not a prime number

or we may use some special mathematical notation, like

a≠0 or a≮b

when we want to say that a=0 is not true or that a<b is not true.

Using English as an internal representation in a computer program brings its own problems, not least of which is the essential ambiguity of English statements. So we will want to avoid the English formulations. It will also be helpful to regularize the various mathematical conventions. We will therefore adopt the negation symbol ~, writing ~p when we mean that statement p is not true. Thus we will write:

~ n is a prime number
~ a=0 and
~ a<b

~, like the other propositional connectives, is *truth functional* , that is the truth of ~p depends only on the truth of p. ~p is false just when p is true and ~p is true just when p is false. We can sum up this relationship with a *semantic tree*, see figure 2-1. The word semantic means 'concerned with meaning'; the semantic tree is a tree which gives the meaning of ~. For those not familiar with mathematical trees, appendix II explains the terminology.

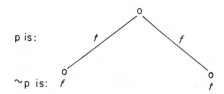

Figure 2-1: Semantic Tree for ~p

This, rather simple, tree has two arcs coming from the root, corresponding to the two possible truth values for p, true and false (abbreviated as *t* and *f*). At the tip of each branch it has the truth value of ~p corresponding to the values assigned to p on that branch. We will meet more complicated trees in the following sections.

2.1.2 Conjunction

Next we will consider how two statements can be connected so that the truth of both of them is asserted. In English, this can be done with words like 'and', 'but', 'where', etc. Consider, for instance,

2<X and X<10
2<X but X≠3
a.X + b = 0 where a≠0

There are also special mathematical conventions like:

$2 < X < 10$

We will replace all of these with the *conjunction* symbol, &. That is, p&q will mean both p and q are true. Again we can make this rather more precise by defining & with a semantic tree, as in figure 2-2.

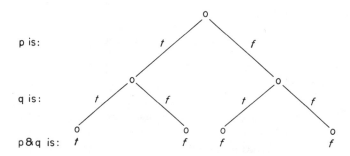

p is:

q is:

p&q is:

Figure 2-2: Semantic Tree for p&q

The truth values given at the tips of the branches are determined by the values assigned to p and q on those branches. Thus if we start from the root and follow the arc where p is true then the arc where q is false, we find that p&q is false.

2.1.3 Disjunction

Just as two statements can be connected to assert the truth of both, they can also be connected to assert the truth of one or the other. This effect can be achieved in English by saying

X≤Y or X≥Y   (i)

or we can use special mathematical conventions like

X = ±2   (ii)

This is called *disjunction*.

As usual we will use a special symbol, v, to indicate disjunction. That is, p v q, will mean that at least one of p and q is true. We will write:

$$X \leqslant Y \text{ v } X \geqslant Y$$
$$X = 2 \text{ v } X = -2$$

The semantic tree for v is shown in figure 2-3.

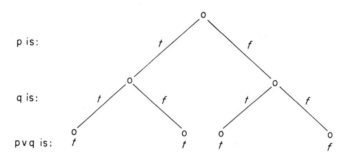

p is:

q is:

pvq is:

Figure 2-3: Semantic Tree for pvq

Notice how we have settled the ambiguous case when both p and q are true. We have made pvq true. This is called *inclusive or*. If we had chosen to make pvq false we would have defined *exclusive or*. Both cases are common in Mathematics (we gave an example of each case in (i) and (ii) above). Inclusive or is usually given precedence in Propositional Logic because it has a nice duality with &. To see this exchange the *t*'s for *f*'s, and vice versa, at the tips of the semantic tree for v and compare the result with the semantic tree for &. Inclusive or is often sufficient to represent disjunctions like (ii) above, because the exclusivity of the cases is implied by the context, e.g. X cannot equal both −2 and +2, since ~2=−2.

Exercise 1: Draw the semantic tree for exclusive or.

2.1.4 Implication

We often want to say that the truth of one statement implies the truth of another. For instance,

If n is an odd number then n is prime
n is prime if n is an odd number
n is prime whenever n is an odd number
For n to be prime it is sufficient for n to be odd
n is an odd number implies n is prime

We will replace these English formulations with the *implication* arrow, →, writing n is an odd number → n is prime

The semantic tree for implication is given in figure 2-4. Notice how we have decided the ambiguous cases where p is false, by making p→q true, regardless of the value of q. This version of implication is called, *material implication*. It is the most common version in Propositional Logic, doubtless because it bears a pretty relationship to v.

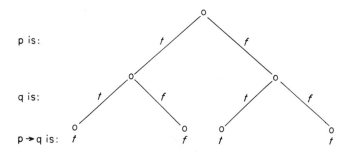

p is:

q is:

p→q is:

Figure 2-4: Semantic Tree for p→q

2.1.5 Double Implication

We often want to assert that not only p→q but also that q→p. This is sometimes done by using the phrase 'if and only if' or its conventional shortening 'iff'. It is also done by phrases like 'a necessary and sufficient condition'.

X+2=3 if and only if X=−1

For a number to be divisible by 15 it is necessary and sufficient that it be divisible by 3 and by 5

We will replace these English formulations with the *double implication* arrow, ⟷.

A note of caution: The word 'if' is often used where double implication is intended, especially in definitions. Consider, for instance,

A number is prime if it has exactly two divisors.

Presumably, this is the only way a number can be prime. The reader is meant to interpret the 'if' as 'iff'.

The semantic tree for double implication is given in figure 2-5.
Notice the duality between this tree and the one for exclusive or which you drew in exercise 1.

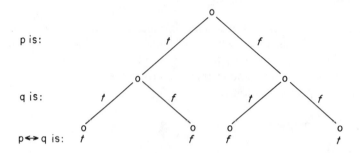

Figure 2-5: Semantic Tree for p⟷q

2.2 Propositional Formulae

Now that we have some connectives: ~, &, v, →, ⟷, we can begin to use them to put together some more complicated statements. Suppose we start with the propositions:

n is an odd number,
n is prime

Then we can use negation to form

~ n is prime

and then disjunction to form:

~ n is an odd number v n is prime

and then conjunction to form:

~ n is an odd number v n is prime & n is an odd number

and so on, and so on.

It will help clear up ambiguities if we establish a precedence ordering among the connectives and use brackets to clear up any remaining conflicts, e.g.

{~ n is an odd number v n is prime} & n is an odd number

We can omit brackets around '~ n is an odd number' because the conventional precedence ordering is that ~ binds tighter than the other connectives, so that ~p v q means (~p) v q rather than ~(p v q). This is similar to the situation in arithmetic where −2 + 3 means (−2) + 3 rather than −(2 + 3).

The precedence order is that ~ binds tightest, & and v bind next tightest, and → and ⟷ bind loosest. Hence,

~p v q	means	(~p) v q
p & q → r v s	means	(p & q) → (r v s)
p & q v r	is ambiguous	

We can summarize this ability to form new formulae from old in the following *recursive* definition. This definition is called *recursive* because it applies to itself, but without getting into an infinite regression.

Definition 1: Formulae

1. A proposition is a formula.
2. If p and q are formulae then the following are also formulae: ~p, p&q, pvq, p→q, p⟷q.
3. Only expressions formed by rules 1 and 2 above are formulae.

It will be convenient to use the letters p, q, r etc, as above, to stand for arbitrary propositions, e.g. {~p v q} & p.

We will sometimes want to represent formulae as trees. The tips of these trees will be labelled by propositions and the other nodes by connectives. The label of the root node is said to be the *dominant* connective. For instance, {~p v q} & p will be represented as

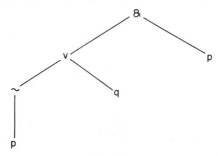

Note how the tree reflects the recursive structure of the formula. Each subformula is represented as a subtree. If a formula is made from a connective and some subformulae then its tree is formed from the subtrees of these subformulae connected by a parent node labelled by the connective.

2.2.1 Semantic Trees

We can draw semantic trees for these more complex formulae, just as we did for the simple ones in section 2.1. Consider the formula:

~(~p v ~q)

The first step is to list the propositions it contains: in this case just p and q. We then build a semantic tree for these sentences, with no labels on the tips.

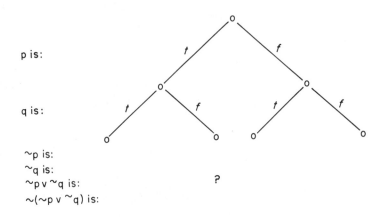

Figure 2-6: Semantic Tree Framework for Two Propositions: p and q

Below this tree we can list the subformulae, of the formula in question, in increasing order of complexity as shown in figure 2-6.

We can then fill in the values of each of these formulae, below each tip, as follows:

- Consider each of the formulae in top/bottom order.
- Consider the dominant connective of the formula. Its parameters will have been assigned values either on the corresponding branch of the tree or by some previous iteration of this process.
- Look up the semantic tree for this connective. Find the branch which assigns these values to its parameters. Find the value at the tip of this semantic tree. This is the value of the formula on the current tip.

Thus, suppose we were trying to fill in the value of ~p v ~q below the second tip (i.e. the slot marked ? in figure 2-6). We assume that the rows for ~p and ~q have already been filled in and the values *f* and *t*, respectively assigned to the second tip. Looking at the semantic tree for v (figure 2-2) we see that when *f* and *t* are assigned to the parameters of v that *t* is assigned to the tip. Thus we replace ? with *t*.

Repeating this process for the remaining formulae and tips gives the tree in figure 2-7. Compare this tree with the tree for p&q. It is the same. p&q and ~(~p v ~q) have the same value for all assignments of truth values to p and q. We say that the two formulae are *logically equivalent* or just equivalent for short.

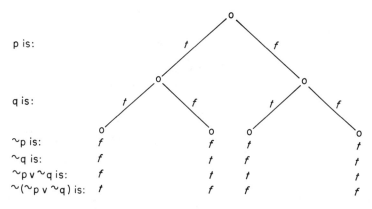

p is:

q is:

~p is:
~q is:
~p v ~q is:
~(~p v ~q) is:

Figure 2-7: Semantic Tree for ~(~p v ~q)

Exercise 2: Draw a semantic tree for the expression ~p v q. Compare this with the semantic tree for p→q.

2.2.2 Equivalences

Using semantic trees we can establish equivalences between different propositional formulae. We have already seen that p&q and ~(~p v ~q) are equivalent. And when doing exercise 2 above you should have noticed that ~p v q and p→q were also equivalent. You may also like to show that p⟷q and (p→q) & (q→p) are also equivalent.

These discoveries should not come as too much of a surprise. Equivalent formulae 'say the same thing'. If you reflect on the meaning of p⟷q {p if and only if q} and (p→q) & (q→p) {p implies q and q implies p} then you will see that they are really two different ways of 'saying the same thing'. Try the same exercise with the other equivalent formulae above until you convince yourself that they really say the same thing too.

The discovery of these equivalences means that there is some redundancy in our connectives. We need not have introduced the connective, ⟷, at all. Whenever we felt the need for it we could have replaced it with the equivalent (p→q) & (q→p). However, this equivalent expression is a bit clumbersome. It is often more convenient to use the shorter, but redundant, ⟷.

In a similar way we could replace all occurrences of p&q with the equivalent ~(~p v ~q) and all occurrences of p→q with the equivalent ~p v q. In this way we can whittle down the connectives we actually need to two, v and ~.

In fact, if we had introduced the connectives, Sheffer stroke and dagger (also known as Nand and Nor), we would have found that either one of them would do, all on its own. Their definitions are:

$$p|q \longleftrightarrow \sim(p \& q)$$
$$p \downarrow q \longleftrightarrow \sim(p \lor q)$$

2.2.3 Tautologies and Contradictions

Consider the formula, $p \lor \sim p$. If we build its semantic tree we will discover that the labels of all its tips are t. That is, it is always true, regardless of the values assigned to its propositions. This is really not surprising. After all it says that either p is true or p is not true. A fairly obvious observation. Such a formula is called a *tautology*.

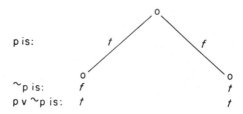

p is:

~p is:

p v ~p is:

Figure 2-8: Semantic Tree for p v ~p

Exercise 3: Show that $\sim\sim p \longleftrightarrow p$ and $(p \rightarrow q) \longleftrightarrow (\sim p \lor q)$ are also tautologies.

As well as formulae whose semantic trees have t at all their tips, we can have formulae whose semantic trees have f at all their tips, i.e. formulae which are false for all assignments. Such a formula is called a *contradiction*

Exercise 4: Show that $p \& \sim p$ is a contradiction.

In exercise 3 you will have shown that a tautology can be formed from a double implication between two of the formulae we showed equivalent in section 2.2.2. This is true in general. If A and B are two equivalent formulae, then A⟷B is a tautology. The reverse is also the case. If A⟷B is a tautology then A and B are equivalent.

Theorem 2: A and B are equivalent formulae if and only if A⟷B is a tautology.

Proof: If A and B are equivalent then, by definition, they have the same truth value for all assignments of truth values to the propositions they contain. Consider any tip of the semantic tree of A⟷B. On the branch above this tip various assignments of truth values will have been made to the propositions in A and B. Substitute these into A and B. The resulting values of A and B will be either both t or both f. In either case the value of A⟷B will be t (see semantic tree for ⟷). Therefore A⟷B is a tautology.

If A⟷B is a tautology then every tip of its semantic tree is labelled *t*. This could only have happened if the assignment on the branch above made A and B either both *t* or both *f* (see semantic tree for ⟷). In either case A and B have the same value. Therefore A and B are equivalent.   QED

2.2.4 Identifying Correct Arguments – Part 1

The apparatus of semantic trees can be used to identify arguments whose correctness only depends on the way the propositions in it are connected. We will call such arguments, *boolean*, after George Boole who first classified such arguments. Correct boolean arguments constitute the theorems of the mathematical theory, *Propositional Logic*. We will have no need to develop this theory, e.g. we do not need to give the axioms of Propositional Logic, since they are not required in order to understand automatic theorem proving.

For instance, consider the argument we met in section 1.3.1.

> Either all reasoning is syllogistic or Aristotle was wrong.
> All reasoning is not syllogistic.
> ───   (i)
> Therefore, Aristotle was wrong.

This contains only two constituent propositions, 'all reasoning is syllogistic' and 'Aristotle was wrong', connected by v, ~, an implicit & between the two hypotheses and a → between the hypotheses and the conclusion. Hence we can formalize it as

> [(All reasoning is syllogistic v Aristotle was wrong) &   (ii)
>   ~ All reasoning is syllogistic] → Aristotle was wrong

The semantic tree for this formula is given in figure 2-9, from which we can

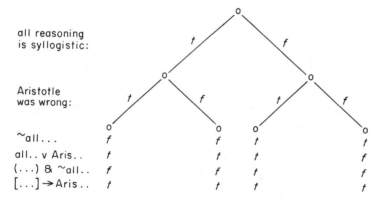

Figure 2-9: Semantic Tree for Formula

see that it is a tautology, i.e. its truth is independent of the truth of its constituent propositions. Hence it is a correct argument form.

The technique exemplified above can easily be implemented in a computer program which, given a formula in logical notation, e.g. (ii) above, could extract the constituent propositions, build a semantic tree and check to see if the formula is a tautology. Such a computer program would be a *decision procedure* that is a procedure which, applied to an argument, is guaranteed to stop after a while and say whether the argument is correct or not. Any area of Mathematics, like boolean arguments, for which there is a decision procedure, is called *decidable*. In subsequent chapters we will discover areas of Mathematics for which, not only do we not have a decision procedure, but we could not ever have one. Such areas are called *undecidable*.

It is even possible to write a computer program to translate the mathematical English formulation of the argument, e.g.(i) to the logical formulation, e.g.(ii), but for this we must wait until chapter 14.

Exercise 5: Show that 'modus ponens', i.e.

P implies Q

P
_____

Therefore Q

is a correct argument form.

How about the other argument forms in section 1.3.1? Can we identify these as correct too? Unfortunately, these rely on the internal structure of the propositions, i.e. they are non-boolean, and we will need the apparatus of the next chapter to deal with them.

2.3 Summary

In this chapter we have introduced the truth functional connectives: negation, conjunction, disjunction, implication and double implication, and defined propositional formulae, which are composed of propositions joined together by connectives. The meaning of these formulae can be calculated by constructing their semantic trees. Formulae which are true on all branches of their semantic tree are called tautologies, and those which are false on all branches are called contradictions. We can test whether a boolean argument is correct by seeing whether it translates into a tautology.

3

The Internal Structure of Propositions

☐ This chapter is an introduction to Predicate Logic.
☐ Section 3.1 introduces the primitive parts of the proposition: the functions, predicates, variables and constants.
☐ Section 3.2 introduces the quantifiers.
☐ Section 3.3 considers formulae made from these primitive parts and defines the meaning of such formulae with the aid of interpretations.
☐ Section 3.4 uses interpretations to identify correct argument forms.

We have some machinery for representing the way in which propositions can be combined and a mechanical method, using semantic trees, for deciding when a formula is always true (a tautology) just by virtue of its structure and, hence, for deciding the correctness of boolean arguments. But we are still representing the propositions themselves in a mathematical English, e.g. 'n is prime'. We now turn our attention to representing the internal structure of propositions. This will lead us to extend Propositional Logic to Predicate Logic, and to a method for deciding the correctness of non-boolean arguments.

3.1 Functions and Predicates: Variables and Constants

We can draw some inspiration here from the existing practice in mathematics. A typical mathematical proposition is

$$\sin(90-X) = \cos(X) \qquad \text{(i)}$$

In this proposition we can identify several different kinds of beast.
- First there is the *constant*, 90. In this case a number which represents one of the objects about which algebraic laws, like (i), above express some truth.
- Then there is the variable, X. This too represents a number, but not a fixed one like 90. X here stands for any number.
- Next are the functions, sin, cos and $-$. These take in numbers and return other numbers, e.g. sin takes 90 and returns 1; given 30, instead, it would return 1/2.
- Lastly is the *predicate*, $=$. This is similar to the functions, but instead of taking in numbers and returning numbers, it takes in numbers and returns a truth value, true or false.

We will adopt a convention that words starting with a capital letter (sometimes just a single capital letter as here) stand for variables, whereas words starting with a lower case letter stand for constants. This will enable us to use a much wider range of symbols for variables than the normal, X, Y, Z etc.

Each proposition contains only one predicate, but it may contain any number of constants, variables and functions. Functions are applied to constants and variables to form *terms*, e.g. $90 - X$. 90 and X are called the *parameters* of $-$. These terms can then, in their turn, be used as the parameters of further functions to form new terms, e.g. $\sin(90-X)$. Finally the terms are used as the parameters of a predicate to form a proposition, e.g. $\sin(90-X)=\cos(X)$.

Some functions and predicates take only one parameter, e.g. sin and cos. They are called *unary*. Some take two parameters, e.g. $-$ and $=$. They are called *binary*. Some take three parameters, e.g. mod. They are called *ternary*. More generally a function or predicate which takes n parameters is called *n-ary*. The number of parameters a function or predicate takes is called its *arity*. Unary predicates are sometimes called *properties* and non-unary predicates *relations*. Nullary functions and predicates are also allowed, i.e. those which take no parameters. In fact we have already met them both: The nullary functions are just the constants and the nullary predicates are the truth values, t and f.

Table 3-1 contains some standard functions and predicates of different aritys (applied to variable parameters to aid clarity).

Table 3-1: Standard Functions and Predicates of Different Aritys

Arity	Functions	Predicates
0	13	t
	e	f
1	$\sin(X)$	
	$\cos(X)$	
2	$X+Y$	$X=Y$
	$X-Y$	$X<Y$
	X^Y	$X\|Y$
	$\log_X Y$	
3		$X=Y \bmod Z$

When we want to represent an arbitrary function or predicate of arity n, applied to some parameters we will use the notation

$f(t_1,\ldots,t_n)$

and

$p(t_1,\ldots,t_n)$

where the t_i are arbitrary terms.

When the t_i are of only peripheral interest we will sometimes use bold face letters to indicate a vector of n parameters, e.g.

f(**t**) and p(**t**)

This notation, where the function or predicate symbol is followed by its parameters contained in parenthesis, is called *Functional Form*. As can be seen in table 3-1, it is rarely used in practice. Symbols are often *infixed* between the parameters, e.g. X+Y. Parameters appear as sub– or super-scripts to symbols and even to other parameters. No holds are barred! We will generally stick to the established mathematical practice, otherwise this book runs the risk of becoming more obscure than necessary, with expressions like $=(+(2,2),4)$.

Some alternative notations to Functional Form are discussed in appendix III.

Armed with all this terminology we are in a position to make proper (recursive) definitions of terms and propositions.

Definition 1: Terms and Propositions

1. A variable or a constant is a term.
2. If $t_1,\ldots,t_n$ are terms, f is an n-ary function and p is an n-ary predicate then

   $f(t_1,\ldots,t_n)$ is a term
   and
   $p(t_1,\ldots,t_n)$ is a proposition

3. Only expressions formed using rules 1 and 2 above are terms or propositions.

Our representation of formulae as expression trees can be extended to propositions and terms. For instance, the expression $\sin(90-X) = \cos(X)$ can be represented by the tree in figure 3-1.

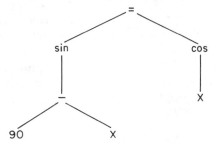

Figure 3-1: The Representation of sin(90−X) = cos(X) as an Expression Tree

Here the tips are labelled by constants or variables and the other nodes by functions or predicates.

Where does all this leave our initial example proposition, 'n is prime'? Since this expresses a proposition which may either be true or false it must contain a predicate. Let us invent a new predicate, is-prime. Let it be a unary predicate, which takes in a number and returns true if and only if that number is prime. Then our proposition can be represented by

is-prime(N)

where n is represented as the variable N. (Should n be a variable or a constant? Does 'n is prime' assert all numbers are prime or only one particular number?)

3.2 The Status of Variables

Variables in mathematical expressions often have an ambiguous status, whose resolution depends on the context. For instance, the X in

$$(X+Y).(X-Y) = X^2 - Y^2 \qquad \text{(ii)}$$

is usually intended to stand for any X and similarly for Y. We will call this the *universal* usage. However, the X in

$$\text{Solve} \quad X^2 + 2.X + 1 = 0 \quad \text{for} \quad X \qquad \text{(iii)}$$

stands for some particular number, whose exact value is not yet revealed. We will call this the *existential* usage.

Mathematicians sometimes try to resolve this ambiguity by calling the first equation an identity and using the predicate $\equiv$ instead of $=$. But this will not do because there are equations in which different variables are to be interpreted in different ways. Consider, for instance,

$$\text{Solve} \quad A.X^2 + B.X + C = 0 \quad \text{for} \quad X \qquad \text{(iv)}$$

Here the A, B and C stand for any number, whereas the X stands for some particular number, whose precise value will depend on A, B and C. We will resolve this ambiguity by announcing the status of each variable with the aid of a *quantifier*, either a *universal quantifier* or an *existential quantifier*. A universal quantifier consists of an upside-down A followed by the variable whose status is being announced, e.g. $\forall X$. An existential quantifier consists of a back-to-front E followed by the variable, e.g. $\exists X$. Thus if we want to assert that (ii) holds for all X and Y we will write

$$\forall X \, \forall Y \quad (X+Y).(X-Y) = X^2 - Y^2$$

On the other hand, if we want to assert that (iii) has a solution we will write

$\exists X \quad X^2 + 2.X + 1 = 0$

In the case of (iv) the precise value of X depends on A, B and C and this can be represented by the order of the quantifiers

$\forall A \; \forall B \; \forall C \exists X \quad A.X^2 + B.X + C = 0$

Putting the existential quantifier for X before the universal quantifiers would imply that all quadratic equations had the *same* solution.

Some of the most subtle questions of variable status occur in Analysis. Consider the assertion that $1/X$ tends to ∞ as X tends to 0. This is usually expressed in English as

> For all M, there exists a Delta such that $|X| \leqslant$ Delta implies that $|1/X| > M$

Expressed in our notation this becomes

$\forall M \exists \text{Delta} \forall X \quad |X| \leqslant \text{Delta} \to |1/X| > M$ \hfill (v)

Note how (as implied by the definition) Delta depends on M, but not on X.

Exercise 6: Consider the following assertions that e^x is continuous and *uniformly continuous*. Express each of them in our formal notation. Compare the resulting formulae.

> *Continuous* – At each point x, for any positive ϵ there exists a δ such that $|e^y - e^x| < \epsilon$ whenever $|y-x| < \delta$.

> *Uniformly Continuous* – For any positive ϵ there exists a δ such that $|e^y - e^x| < \epsilon$ whenever $|y-x| < \delta$.

In section 2.2 we gave a definition of 'formula'. This must now be updated to include the quantifiers. The first step is to retract clause 3 of the definition, which restricted the formulae to those defined using the connectives – we are going to add a new way to form formulae, using quantifiers. For the sequel we will find it useful to pay attention to the variables whose status has not been announced with quantifiers; the *free variables*. This entails redoing the definitions of terms and propositions giving the set of free variables they contain. We use the standard curly bracket notation for sets, e.g. the set containing just X, Y and Z is written $\{X,Y,Z\}$.

Definition 2: Terms, Propositions, Formulae and Sentences.

1. A constant is a term with free variables $\{\}$.
2. A variable, X, is a term with free variables $\{X\}$.
3. If **t** is a vector of n terms with free variables **vars**, v is the union of these

variables, f is an n-ary function and p is an n-ary predicate then

f(**t**) is a term with free variables v

and

p(**t**) is a proposition with free variables v.

4. Only expressions formed using rules 1, 2 and 3 above are terms or propositions.
5. A proposition with free variables v is a formula with free variables v.
6. If A and B are formulae with free variables va and vb then ~A is a formula with free variables va and A&B, AvB, A→B and A⟷B are formulae with free variables va ∪ vb
7. If A is a formula with free variables v containing X then ∀X A and ∃X A are formulae with free variables v minus X.
8. Only expressions formed by rules 5, 6 and 7 are formulae.
9. A formula with no free variables is a *sentence*.

The sentences, or formulae with no free variables, are of especial interest to us because they are the only formulae in which the statuses of the variables are unambiguous and thus are the only formulae which can be unambiguously assigned a truth value, *t* or *f*. Even though the status of a free variable is ambiguous, the universal status is the preferred reading. The sentence obtained from a formula by attaching a universal quantifier to it for each of its free variables is called its *closure*. In subsequent chapters we will often use an unquantified formula as an abbreviation for its closure.

3.3 The Meaning of Formulae

When is a sentence containing quantified variables true and when false? If symbols like =, 0, etc are interpreted in the conventional way then it is clear that

∀X X=X is true
∀X X=0 is false
∃X X=0 is true
and ∃X   X=X+1 is false

For instance, ∀X X=X is true because 0=0, 1=1, 2=2, etc are true. ∃X X=0 is true because if X were 0 then X=0 is true. A formula A' formed by substituting objects from a set, e.g. the natural numbers, for all free variables in A is called an *instance* of A, e.g. 2+2=4 and 2+2=5 are both variable free instances of X+Y=Z. Our intuitive interpretation of the meaning of the quantified formulae above used the rules,

– ∀X A is true iff every instance of A is true.

– ∃X A is true iff some instance of A is true.

From the semantic trees in section 2.1 we can deduce similar rules for the connectives, e.g.

A&B is true iff A is true and B is true.

Thus the truth value of a sentence can be recursively reduced to the truth value of each member of a collection (possibly infinite) of propositions not containing variables.

The meaning of formulae, which are not sentences, can be defined as follows. A formula will be said to be true iff all its instances are true (i.e. iff its closure is true). A formula will be said to be false iff all its instances are false . This leaves a lot of unsettled cases with some true instances and some false ones. In particular, note that a formula which is not true is not necessarily false and that a formula which is not false is not necessarily true.

We will find it useful to be able to manipulate substitutions as objects. A *substitution* is a set of pairs, where each pair is a variable, X, and the term, t, substituted for it. If we represent such a pair by t/X then the substitutions which form 2+2=4 and 2+2=5 from X=Y, can be denoted by {2+2/X, 4/Y} and {2+2/X, 5/Y}. We can read t/X as 't is substituted for X' or 'X is bound to t'. We will let Greek letters like, ϕ, Θ and ψ stand for arbitrary substitutions. We will denote the result of applying a substitution, ϕ, to the formula, A, by writing the substitution immediately after the formula, i.e. Aϕ, Using this notation

$$(X=Y)\{2+2/X, 4/Y\} \equiv (2+2=4)$$

where $\equiv$ means "is identical to".

Note that bound variables are immune from substitution. For instance, substituting 0 for X in $\forall$X X=X causes no change.

$$(\forall X \ X=X)\{0/X\} \equiv (\forall X \ X=X)$$

Recall that an instance is formed by substituting objects from a set for all free variables in a formula e.g. the numbers 1,2,3 etc. and not arbitrary terms like 2+2, 2+X, etc.

3.3.1 Interpretations

This discussion leaves an open question

What objects must we consider substituting for the variable X in A when calculating the truth value of $\forall$X A or $\exists$X A.

The truth value of the formula may hang on our choice. Consider the formula

$$\forall X \ \exists Y \ Y+1 = X$$

This is true over the integers, but false over the natural numbers. That is, if we consider substituting integers for the variables, X and Y, then whatever integer is substituted for X, the number 1 less can be substituted for Y to make a true instance. However, if we are not allowed negative numbers then there is no number to substitute for Y to make $Y+1 = 0$ true. Similarly, the formula

$$\forall X \forall Y \{ X<Y \rightarrow \exists Z (X<Z \& Z<Y) \}$$

is true over the reals, but false over the integers.

We need a relative notion of truth rather than an absolute one. In future we will talk about truth in an *interpretation*, where the interpretation is a system which provides

(a) A set of objects for the variables to range over. For instance, this might be the set of natural numbers, integers or reals, or it might be a finite set of objects. We will call it the *universe* of the interpretation.
(b) An assignment of meaning to the functions and predicates that appear in the formula, e.g. $+$, $=$, etc. We will want to be able to specify the conventional meanings we appealed to informally above, and some unconventional meanings.

Part (a) of the interpretation is provided by naming a non-empty set of objects as the universe, e.g. the natural numbers, $\{0,1,2,3,4,5,...\}$.

Part (b) of the interpretation is provided by associating *calculation procedures* with the constants, functions and predicates of the formula.

   – The calculation procedure associated with a constant is trivial; it consists of assigning it a member of the universe as its *value*. We will often construct universes from the constants of a theory and assign a constant to itself as its own value.
   – The calculation procedure associated with a function should specify the result of applying the function to all combinations of parameters drawn from the universe. This result will be another object in the universe.
   – The calculation procedures associated with a predicate should specify the truth value obtained by applying the predicate to all combinations of parameters drawn from the universe.

One way of specifying these calculation procedures is to give tables like those in figure 3-2. Another way is to give a computer program.

Note that the tables defining each function and predicate give one and only one result for each combination of parameters from the universe. Thus this method of interpreting formula does not allow the possibility of multiply defined functions, e.g.

* $\sqrt{4}=2$ and $\sqrt{4}=-2$

or ambiguous predicates, nor does it allow the possibility of functions and predicates undefined in some region, e.g. 2/0 must be assigned some default value. The assignment of a default value to normally undefined regions of functions and predicates can be something of a practical nuisance.

3.3.2 Interpreting Formulae

For instance, consider the formula

$$\forall X \, \exists Y \quad X=Y \lor X=Y+1$$

To keep matters simple we will use an interpretation with the finite universe, $\{0,1\}$, and assign the constants 0 and 1 to themselves. The formula includes the binary function, $+$, so we must specify the results of applying $+$ to 0 and 1 in all possible ways, and these results must always be either 0 or 1. One way of doing this is given by the addition table in figure 3-2. The formula also includes the binary predicate, $=$, so we must specify the truth value obtained by applying $=$ to 0 and 1 in all possible ways. A way of doing this is also given in figure 3-2. We will call the interpretation so defined *boole*.

+	0	1		=	0	1
0	0	1		0	*t*	*f*
1	1	0		1	*f*	*t*

Figure 3-2: Calculation Procedures for $+$ and $=$

The meaning of our formula is determined by boole.

$\forall X \, \exists Y \, X=Y \lor X=Y+1$ is true in boole
    iff (by meaning of $\forall$)
$\exists Y \, 0=Y \lor 0=Y+1$ is true in boole and
$\exists Y \, 1=Y \lor 1=Y+1$ is true in boole
    iff (by meaning of $\exists$)
either $0=0 \lor 0=0+1$ or $0=1 \lor 0=1+1$ is true in boole and
either $1=0 \lor 1=0+1$ or $1=1 \lor 1=1+1$ is true in boole
    iff (by semantic tree for $\lor$)
either $0=0, 0=0+1, 0=1$ or $0=1+1$ is true in boole and
either $1=0, 1=0+1, 1=1$ or $1=1+1$ is true in boole
    iff (by table for $+$)
either $0=0, 0=1, 0=1$ or $0=0$ is true in boole and
either $1=0, 1=1, 1=1$ or $1=0$ is true in boole
    iff (by table for $=$)
either *t*, *f*, *f* or *t* is true in boole and

either f, t, t or f is true in boole
  iff (by meaning of English)
t

When we design some axioms to capture a mathematical theory, for which there are established calculation procedures, say the theory of natural numbers, then we will want all the axioms to be true in the interpretation defined by these procedures. It will be convenient to be able to refer to this standard interpretation. We will call it *arith*. That is, arith is the interpretation with the natural numbers as universe and the standard arithmetic calculation procedures associated with = (equality), . (multiplication) and + (addition).

3.3.3 Some Definitions

When a formula is true in an interpretation the interpretation is said to be a model of the formula. This notion can be extended to a set of axioms by considering the formula made by conjoining all the axioms together (taking care not to get any accidental coincidences of variables). Thus a set of axioms can have a model. An axiomatization of the arithmetic of the natural numbers would have arith as a model.

Consider how each of the interpretations of a sentence might treat it. The following cases can arise.
1. If *all* the interpretations are models of the sentence then we say that the sentence is *logically valid*. Otherwise, the sentence is *logically invalid*.
2. If *none* of the interpretations are models of the sentence then we say the sentence is *unsatisfiable*. Otherwise, the sentence is *satisfiable*.

Note that a sentence can be satisfiable without being valid and invalid without being unsatisfiable. For example,

- $\forall X\ X=0 \rightarrow X=0$ is logically valid, because all its instances are tautologies.

- $\forall X\ X=0$ is logically invalid, because arith is not a model.

- $\forall X\ X=0\ \&\ \sim X=0$ is unsatisfiable, because all its instances are contradictions.

- $\forall X\ X=X$ is satisfiable, since arith is a model.

The notion of a logically valid sentence extends the notion of a tautology, that we met in section 3. In fact, the logically valid sentences constitute the theorems of the mathematical theory, *Predicate Logic*, also known as *First Order Logic*. Similarly, the notion of an unsatisfiable sentence extends the notion of a contradiction. We establish this connection more formally in chapter 16.

3.4 Identifying Correct Arguments – Part 2

Can we now use the extended definition of the meaning of a formula, to deal with the remaining argument forms of section 1.3 – the non-boolean ones? Consider, for instance, the substitution rule:

$$\frac{A(X)}{A(T)} \tag{vi}$$

The first thing to note is that this does *not* mean the same thing as

$$A(X) \rightarrow A(T) \tag{vii}$$

In fact (vii), while plausible at first sight, is not logically valid: in fact it is not true even in the standard interpretation, arith. This may be clearer in a particular case. Let $A(X)$ be $X=0$ and T be 1 then (vii) is

$$X=0 \rightarrow 1=0$$

However, one of the variable free instances of this is

$$0=0 \rightarrow 1=0$$

under the substitution $\{0/X\}$; and this is false in arith, since arith assigns $0=0$ to t and $1=0$ to f and $t \rightarrow f$ is f.

(vi), however, does not suffer this defect. It asserts that $1=0$ can be deduced from $X=0$, which is fair enough since $X=0$ means 'all numbers are equal to 0'. It does not allow us to deduce $1=0$ from $0=0$.

This means that we cannot represent arguments as implications between hypothesis and conclusion and test for tautologyhood, as we did in section 2.2.4. Instead we must interpret the hypothesis and conclusion separately. We will use the following criterion

Definition 3: An argument

$$\frac{H}{C}$$

is correct iff every interpretation of H and C which is a model of H is also a model of C. We will say that C is a *logical consequence* of H.

If, in addition, H is a logical consequence of C, we will say that they are *logically equivalent*. This extends the definition of section 2.2.2 to formulae with variables.

Consider how we may use this definition to show the correctness of the substitution rule

$$\frac{A(X)}{A(T)}$$

There is a caveat to attach to this rule; T should not contain a free variable which is bound in A(X). For instance, we are not allowed to substitute Y+1 for X in $\exists$Y Y=X, giving $\exists$Y Y=Y+1.

Let M be a typical model of A(X) (which is also an interpretation of A(T)). We must show that M is also a model of A(T), i.e. that A(T) is true in M. Now every variable free instance of A(T) is also a variable free instance of A(X) and hence true in M. Therefore A(T) is true in M. QED.

In a similar manner we can show the correctness of the other arguments from section 1.3.1. Consider, for instance,

> This is an argument.
> This is not propositional.
> This is not syllogistic.
> _____
>
> Therefore, some arguments are neither propositional nor syllogistic.

   This can be formalized as

argument(this) &
~propositional(this) &
~ syllogistic(this)
_____
$\exists$X argument(X) & ~[propositional(X) v syllogistic(X)]

The universe of any interpretation of the hypothesis contains the object, 'this', hence the calculation procedures for the predicates must assign meanings to the propositions:

argument(this),
propositional(this) and
syllogistic(this)

Any model of the hypothesis must assign the first of these to t and the second and third to f. Thus, by the semantic trees for v, ~ and &, the instance of the conclusion defined by the substitution

{this/X}

will be true and hence the conclusion will be true in all models of the hypothesis. QED.

Exercise 7: Show the Darii syllogism of section 1.3.1 is a correct argument form.

Exercise 8: Show that $\exists X \sim A(X)$ is a logical consequence of $\sim\forall X\, A(X)$.

The technique illustrated above, for testing the correctness of non-boolean arguments, cannot be as easily translated into a computer program as the boolean technique of section 2.2.4 could. We cannot always test all models of a hypothesis because there may be infinitely many of them and testing the conclusion may involve considering an infinite number of instances. Thus the technique does not constitute a decision procedure, because it may not terminate; it may grind on forever without giving an answer. It can be shown that non-boolean arguments constitute an essentially *undecidable* area of Mathematics, i.e. there could not be a decision procedure.

The arguments above either deal with finite cases or exploit some special trick. The best we can do in general is to design a computer program which is guaranteed to terminate only if the argument is correct. This is done by trying to show that the theory consisting of the hypothesis and the negation of the conclusion is unsatisfiable, i.e. has no models. If the conclusion is not a logical consequence of the hypothesis then the model testing may end with a counterexample, but it may go on forever, inconclusively. Fortunately, if the conclusion *is* a logical consequence of the hypothesis this eventually becomes apparent. This is achieved by doing the model testing by a process of deducing theorems of the theory; the deduction of *f* demonstrates that the theory is unsatisfiable. Programs, which are guaranteed to terminate only if the argument is correct, are called *semi-decision procedures*. In chapter 5 we look at how theorems of a theory can be deduced, and in chapter 6 we investigate semi-decision procedures for logical consequence.

3.5 Summary

In this chapter we have looked inside the proposition and defined its internal structure. We found constants, variables, functions and predicates, and gave formal definitions of how these parts could be combined to form terms and propositions. We considered the status of variables, and showed how to distinguish the cases that arise, by using quantifiers. This new notation was given meaning by introducing the idea of an interpretation.

We considered non-boolean arguments, which relied on the internal structure of the propositions. Such arguments were defined to be correct when the conclusion was a logical consequence of the hypothesis. This definition was used to show some arguments correct, but in general the procedure to do this is not a decision procedure; it is not guaranteed to terminate.

4
Miscellaneous Topics

- [ ] This chapter consists of three loosely related sections.
- [ ] Section 4.1 is an introduction to higher order logics, especially typed lambda calculus.
- [ ] Section 4.2 gives the axioms for 3 mathematical theories: equality 4.2.1, group theory 4.2.2 and Peano Arithmetic 4.2.3.
- [ ] Section 4.3 is a series of practical hints on how to represent knowledge using mathematical logic.

4.1 Higher Order Logics

The logical notation introduced so far is that of the Predicate Logic of Frege, sometimes called First Order Logic. This will be enough for most of our purposes, but occasionally we will require something more.

4.1.1 Variable Functions and Predicates

Suppose we want to assert that if two objects, X and Y, are equal then they have the same properties. We will want to use a variable predicate, P, to stand for any property.

$\forall P \ \forall X \ \forall Y \quad X=Y \rightarrow \{P(X) \longleftrightarrow P(Y)\}$

Or suppose we want to give the general definition for a function to tend to ∞ as X tends to 0, rather than just assert that a particular function, like 1/X, does so, i.e. we want to define $limit(F,0)=\infty$ where F is a variable function.

$\forall F \quad limit(F,0)=\infty \longleftrightarrow$
$\quad \forall M \ \exists Delta \ \forall X \quad |X| \leqslant Delta \rightarrow F(X) > M$

First Order Logic only allows variables ranging over objects, e.g. X ranging over 90, e, etc. A logic in which variable functions and predicates are allowed and can be quantified over is called *Second Order*.

4.1.2 Functionals

Where there is a first and second, there must be a third. Consider the integral expression

$\int \cos(X) \, dX$

What sort of beast is the integral sign, $\int$? Clearly it is not a constant, variable or predicate. Is it a function? If it is, what are its parameters? Well it is dependent on its integrand, so cos(X) must be one parameter, and we must know what the integrand is to be integrated with respect to, so X must be another parameter. Hence we could standardize the integral notation to:

$$\int (\cos(X),X)$$

Unfortunately, this will not do. The X above is not really a free variable. For instance, we cannot substitute 90 for X and get a meaningful expression.

$$\# \int \cos(90) \, d90$$

(A # sign in front of an expression indicates that the expression is illegally formed.) We may also change the variable of integration without changing the value of the expression.

$$\int \cos(X) \, dX = \int \cos(Y) \, dY$$

One way out of this difficulty is remove the variable of integration altogether, and make $\int$ a function of a function, e.g. $\int$ takes cos to sin.

$$\int (\cos) = \sin$$

Such a function is called a *functional*. Allowing variable functionals and quantification over them takes us into *Third Order Logic*.

The same problems occur with differentiation and can be cured with the same device, by making d/dX a functional, taking, say, sin to cos.

4.1.3 Lambda Abstraction

Functionals apply to functions, rather than to terms, e.g. to sin rather than to sin(X). This is all very well when we have a convenient symbol to apply it to, but what does $\int$ apply to in the integral

$$\int \sin(X).e^X \, dX \, ?$$ (i)

We do not want to be constantly introducing new functions, e.g.

$$foo(X) = \sin(X).e^X$$

just so we can replace (i) above by

$$\int foo$$

A solution to this is the notion of *lambda abstraction*. We can turn an arbitrary term into the corresponding function by the following device.

$$\lambda \, X(\sin(X).e^x) \text{ is foo}$$

So now (i) above can be written as:

$$\int \lambda X (\sin(X).e^X)$$

By the way, did you spot the deliberate mistake above? Of course, $\int$ does not take cos and return sin, it returns $\lambda X (\sin(X) + c)$ where c is a new constant!

4.1.4 Omega Order Logic

Now we can iterate the process, allowing functions of functionals *(Fourth Order)*, functions of functions of functionals *(Fifth Order)* and so on. Allowing all of these then takes us to *Omega Order Logic*.

A nice way of capturing all the sorts of functions, functionals etc is provided in the *Typed Lambda Calculus* [Church 40], which is one version of Omega Order Logic. Each of the constants is assigned a type. For instance, 90, e and 1/2, may all be assigned type *real*, for real number. and *t* and *f* may be assigned the type *truth*, for truth value.

- Unary functions, e.g. sin, are then mappings from type real to real. If we use the arrow, $\mapsto$ to represent mappings, then this can be summarized as real $\mapsto$ real.
- Binary functions, e.g. +, have type real $\times$ real $\mapsto$ real.
- The functional $\int$ has type (real $\mapsto$ real) $\mapsto$ (real $\mapsto$ real), that is, unary function to unary function.
- Binary predicates, e.g. =, have type real $\times$ real $\mapsto$ truth.
- Binary connectives (e.g. &) have type truth $\times$ truth $\mapsto$ truth.
- $\lambda X \sin(X).e^X$ has type real $\mapsto$ real, since X has type real and $\sin(X).e^X$ has type real.

Thus all the various expressions we have met so far: formulae, propositions, terms, functions, predicates, connectives, variables and constants can be defined in a uniform manner using the terminology of the Lambda Calculus.

Definition 1: The Expressions

1. If s is an expression of type T containing variables, $X_1,...,X_n$, of types, $T_1,..., T_n$, respectively then $\lambda X_1 ... \lambda X_n$ s is an expression of type $T_1 \times...\times T_n \mapsto T$.

2. If f is an expression of type $T_1 \times ... \times T_n \mapsto T$ and $s_1,...,s_n$ are expressions of type $T_1,..., T_n$, respectively, then $f(s_1,...,s_n)$ is an expression of type T.

3. If p is an expression of type $T_1 \times ... \times T_n \mapsto$ truth and Y is a variable of type T_i then $\forall Y$ p and $\exists Y$ p are expressions of type $T_1 \times ... \times T_{i-1} \times T_{i+1} \times ... \times T_n \mapsto$ truth

4. Expressions can only be formed from constants and variables and by application of rules 1, 2 and 3 above.

4.2 Mathematical Theories

Now that we have developed our notation we can achieve our original aim of representing mathematical knowledge in a form suitable for manipulation by a computer program. We will give some sample sets of axioms, so that later we may consider how theorems may be proved from them.

4.2.1 Equality

= is a predicate which crops up throughout mathematics, not just in arithmetic and algebra, but in group theory, set theory etc. Thus the axioms which define =, form a subset of the axioms of all these theories.

Each of the axioms of equality is sufficiently famous to have earned itself a name. The first is *reflexivity* – an object is always equal to itself.

$$X=X$$

The second is *symmetry* – it does not matter how the parameters of = are ordered.

$$X=Y \rightarrow Y=X$$

The third is *transitivity* – that equality is inherited.

$$X=Y \,\&\, Y=Z \rightarrow X=Z$$

The fourth and fifth are very similar and share the name *substitution axiom*. They assert that having equal parameters ensures equal results.

$$X_1 =Y_1 \,\&...\&\, X_n =Y_n \rightarrow f(X_1,...,X_n)=f(Y_1,...,Y_n)$$
$$X_1 =Y_1 \,\&...\&\, X_n=Y_n \,\&\, p(X_1,...,X_n) \rightarrow p(Y_1,...Y_n)$$

In fact, these are *axiom schemata* – we need a version for each n-ary function, f, and each n-ary predicate, p, in our theory. Hence we will need different versions of the fourth and fifth axioms in different theories.

These substitution *axioms* are not to be confused with the substitution *rule*, which allows the substitution of terms for variables in formulae.

> *Exercise 9:* How many of the equality axioms are true for the inequality predicates, $\geq$ and $>$, between real numbers? What additional axioms are needed for inequality over those needed for equality?

Note that these axioms guarantee that, as mentioned in section 3.3.1, functions are defined everywhere and are not multiply defined. We will call these properties of functions *existence* and *uniqueness*, respectively. For instance, if f is a unary function, its existence property is guaranteed by the theorem

$$\exists Y\, f(X)=Y$$

which is easily proved from reflexivity (let Y be f(X)), and its uniqueness property is guaranteed by the theorem

$$f(X)=Y\ \&\ f(X)=Z \rightarrow Y=Z$$

which is easily proved from symmetry and transitivity.

4.2.2 Group Theory

The group axioms are another important constituent of mathematical theories. To start with we will need the equality axioms, reflexivity, symmetry, transitivity and some versions of substitution – just which versions we consider below.

In addition, we will need the axioms to define an identity element, e,

$$X = e \circ X$$
$$X = X \circ e$$
$$\text{where o is a binary function.}$$

and some axioms to assert that every element, X, has an inverse, i(X),

$$X \circ i(X) = e$$
$$i(X) \circ X = e$$

and the *associativity* axiom for o.

$$X \circ (Y \circ Z) = (X \circ Y) \circ Z$$

The functions which occur above are e, i and o, and the only predicate is =. Since e is nullary its substitution axiom is just an instance of the reflexivity axiom, and so can be omitted. The substitution axiom for = is a simple consequence of transitivity, symmetry and reflexivity and can also be omitted. The remaining substitution axioms are:

$$X_1 = Y_1 \rightarrow i(X_1) = i(Y_1) \text{ and}$$
$$X_1 = Y_1\ \&\ X_2 = Y_2 \rightarrow X_1 \circ X_2 = Y_1 \circ Y_2$$

4.2.3 Natural Number Arithmetic

The last in our series of classic sets of axioms are the *Peano axioms* for the Arithmetic of the natural numbers, 0, 1, 2, 3, ... This theory also requires the equality axioms.

The key observation behind the axiomatization is that all the *natural numbers* may be generated by a process of *succession*, starting at 0 and incrementing in 1s. The process of adding a 1 to a number is called forming its *successor* and is denoted by the *successor* function, s, i.e. $s(X)=X+1$.

By using the successor function we can avoid having to have an infinite number of constants in our theory, i.e. 0, 1, 2, 3, ... We have one constant, 0, and represent other numbers by applying s to 0 repeatedly, e.g. 3 is $s(s(s(0)))$.

As with group theory we start with the equality axioms. The first two arithmetic axioms assert that the process of succession will keep generating new numbers: it will never loop back, either to 0

$$\sim 0=s(X)$$

or to some other number

$$s(X)=s(Y)\rightarrow X=Y$$

The next pair of axioms provide a recursive definition of addition in terms of successor. Before giving it let us think about the different kinds of definition we can make in mathematics:

- The simplest kind are *explicit definitions.* The function (or predicate) to be defined appears once, dominating one side of the equation (or double implication), with distinct variables in each parameter position, e.g. $X^2 = X.X$
- A definition which is not explicit is an *implicit definition.* The function may appear on both sides of the equation, or may have non-variable parameters, or may appear in a non-dominating position, etc. Even so, the implicit definition may lay down enough conditions to determine uniquely the function, e.g. the definition of square root over the reals by $(\sqrt{X})^2 = X$ & $\sqrt{X} \geqslant 0$.
- A *recursive definition* is implicit, but in a format which guarantees that the defined function is uniquely determined. It relies on the fact that the objects in the domain can be generated by a process like succession.

The Peano recursive definition of + in terms of s is by two cases: the second argument of + is either 0

$$X+0 = X \tag{ii}$$

or is the successor of some number

$$X + s(Y) = s(X + Y) \tag{iii}$$

In the second case $X + s(Y)$ is defined in terms of $X+Y$. This is, of course, a circular definition, but not a vicious circle, since we can use it to calculate the sum of any two numbers, e.g.

$$s(s(0)) + s(s(0)) = s(s(s(0)) + s(0)) \quad \text{by (iii)}$$
$$= s(s(s(s(0)) + 0)) \text{ by (iii)}$$
$$= s(s(s(s(0)))) \qquad \text{by (ii)}$$

Thus $+$ is uniquely determined by (ii) and (iii), even though it is not explicitly defined.

The third pair of axioms are a recursive definition of multiplication in terms of addition. Again we have the same two cases: the second parameter of . is either 0

$$X.0 = 0 \qquad\qquad\qquad\qquad\qquad\qquad\qquad\qquad \text{(iv)}$$

or is the successor of some number

$$X.s(Y) = X.Y + X \qquad\qquad\qquad\qquad\qquad\qquad \text{(v)}$$

Again $X.s(Y)$ is defined in terms of $X.Y$.

Exercise 10: Calculate $s(0).s(s(0))$ using (iv), (v), (ii) and (iii).

The format of the recursive definitions of $+$ and . was the same:

$$F(X,0,Z) = G(X,Z)$$
$$F(X,s(Y),Z) = H(F(X,Y,Z),X,Y,Z)$$
where Z is a vector of n additional parameters

Except that Z was empty in the definitions of $+$ and . Such a definition is called *primitive recursive* and is a very common type of recursive definition. More general types of recursive definition can be obtained by allowing, for instance, simultaneous recursion on two variables, X and Y, i.e. $F(s(X),s(Y),Z)$ is defined in terms of $F(X,Y,Z)$.

Exercise 11: Give a primitive recursive definition of exponentiation, X^Y, (for natural numbers) in terms of multiplication.

Some versions of natural number arithmetic allow the addition of any number of new axioms provided they conform to the format for explicit or recursive definitions of new functions or predicates, e.g. these theories would allow as new axioms the definition of exponentiation you just designed in the above exercise.

Lastly we need an axiom which says that all the numbers are generated by the succession process. The axiom of mathematical *induction* has essentially this message, because it says that a formula is valid if it is valid for all numbers defined by the succession process.

$$P(0) \ \& \ \forall X \ [P(X) \rightarrow P(s(X))] \rightarrow \forall Y \ P(Y)$$

Note that induction is a second order axiom since P is a variable predicate.

There is a dual relation between induction and recursion which we will make explicit in chapter 11.

That completes our examples of axiom sets for this chapter. We will be considering further axiom sets as they arise in the context of particular Mathematical Reasoning programs. We complete this chapter with some hints on how to build axiom sets.

4.3 Some Practical Hints

Those who intend to build computational models of mathematical (or any other kind of) reasoning are well advised to learn the art of representing knowledge in a logical calculus. One of the main aims of this book is to teach that art and that is why we have invested a disproportionate amount of effort (by the standards of conventional logic textbooks, e.g. [Mendelson 64]) in explaining logical notation. To conclude this teaching we would like to summarize a few practical hints and standard pitfalls.

4.3.1 Function or Predicate?

We have a relationship, r, between some objects, x and y, which we wish to represent. Should we represent it with a function and equality, i.e. $r(x) = y$ or with a predicate, i.e. $r(x,y)$?

This will depend on r. As noted in section 4.2.1, the equality axioms give function notation two properties that predicate notation lacks: the existence of a suitable y given x and the uniqueness of this y, up to equality.

Suppose we want to represent the relationship between a formula, fm, and its proof, prf. Now given fm, prf is neither guaranteed to exist (since fm may not be a theorem) nor is it unique if it does exist (since a theorem may have several proofs). So we cannot represent the relationship by an equation

> \# proof-of(fm) = prf

On the other hand, fm *is* uniquely determined by prf – a proof proves one and only one theorem. So the relationship can be represented by the equation

> theorem-of(prf) = fm

where theorem-of is a function which extracts the theorem proved by a proof, and returns some default value, say 'undef', for terms which are not proofs.

The relationship between a sentence, sent, and a model, m, however, cannot be represented by a function in either direction. A sentence can have several models or none. An interpretation can be a model of several sentences. So we must be contented with a relation, i.e.

is-model(sent,m)

If we must represent an essentially non-function relationship, r, between x and y, with a function, we can always employ the following trick: a function, f_r, is defined from x to the set all y for which r(x,y).

$$f_r(x) = \{y: r(x,y)\}$$

If there are no ys, such that r(x,y) for some x, then $f_r(x)=\{\}$. If there are several then $f_r(x)$ is a non-singleton set.

4.3.2 An Advantage of Avoiding Functions

Despite the advice in the above section there is a strong advantage to be gained if an entire mathematical theory – axioms and theorems – can be expressed without the use of proper functions, but with only predicates, constants, variables and universal quantifiers. In section 16.2, we will see that arguments in such theories are essentially boolean and can be settled by testing a finite number of variable free, quantifier free formulae for tautologyhood using the techniques of propositional logic.

As an illustration, consider the following alternative axioms for group theory. These are modified from the axioms in section 4.2.2 by exchanging expressions of the form, SoT = U and i(S) = T, for, o(S,T,U) and i(S,T), respectively.

identity	o(e,X,X)
	o(X,e,X)
inverse	i(X,Y) $\rightarrow$ o(X,Y,e) & o(Y,X,e)
associativity	o(X,Y,U) & o(U,Z,W) & o(Y,Z,V) $\rightarrow$ o(X,V,W)
	o(Y,Z,V) & o(X,V,W) & o(X,Y,U) $\rightarrow$ o(U,Z,W)

Axioms like these have been used with great success, by George Robinson and Lawrence Wos, as a basis for an automatic theorem prover for group theory, [Wos et al 65]. This success must be attributable, at least in part, to the essential finiteness of the processes involved.

Naturally, there is a price to be paid for such success. The price is that not all theorems of group theory can be proved from these axioms. In particular, it is not possible to prove any theorem which expresses the existence or uniqueness properties of o or i, e.g.

$$\forall X\, \exists Y \quad i\,(X,Y)$$

For a way round this difficulty see chapters 10 and 14.

4.3.3 Variadic Functions and Predicates

There are some functions and predicates which we tend to think of as being able to take any number of parameters – of being of variable arity or *variadic*. This is sometimes expressed by writing a chain of parameters and functions (or predicates), e.g.

$$2 + 3 + 2$$
$$3.a.\sin(x)$$
$$2.3 = 3.2 = 6$$
$$2 \leqslant a \leqslant 5$$

(Note that for this chain notation to be used unambiguously on functions they must be associative. Division, X/Y, is not associative, so 2/3/2 is ambiguous between (2/3)/2 and 2/(3/2).)

Unfortunately, Predicate Logic does not allow variadic functions; the arity of a function must be fixed, and we have fixed the arity of +, ., = and $\leqslant$ at 2.

We can regain something of the flexibility of variadic functions and predicates by the following device. The function or predicate in question is made unary but given a composite parameter; one containing all the original parameters. A *set* of the original parameters is the sort of composite parameter I have in mind, but this is not quite right. Two of the properties of a set are wrong: the elements of a set are unordered and multiple occurrences of an element are merged. The sort of composite parameter we want is a tuple, which is usually called a *list* in computer circles. In a list the elements are ordered and multiple occurrences of the same element are allowed. Lists are delimited with angle brackets and the elements are separated by commas, e.g. <2,3,2>. Our unary (variadic) +, is written as

$$+(<2,3,2>)$$

our unary (variadic) =, as

$$=(<2.3, 3.2, 6>)$$

and our unary (variadic) $\leqslant$, as
$$\leqslant(<2, a, 5>)$$

Although the order of the parameters of functions and predicates is important in general, it is actually immaterial for two of the examples above, + and =, because they are commutative (or symmetric). Thus, it would actually be more convenient to use a composite parameter for such functions which built in commutativity, i.e. in which order was ignored, but multiple occurrences of elements were retained. Such a composite object has been invented in Artificial Intelligence: it is called a *bag*. (If you put two cans of baked beans and a pound of sausages in a shopping bag, they may get jumbled about, but no cans will disappear.) We will delimit bags by square brackets and separate the elements by commas, e.g. [2,3,2]. Our unary (variadic) +, is written as

+([2,3,2])

4.3.4 Representing Negation

Suppose we want to say that a sentence, sent, is unsatisfiable – has no model. A common error is to formalize this as

is-model(sent,none)   (vi)

where 'none' is a special object denoting non-existence. But (vi) means that sent *has* a model called none!

Suppose we had a theorem that f (false) was not deducible from a set of axioms with a model, expressed as:

is-model(Ax,M) $\rightarrow$ ~deducible(Ax,f)

Then substituting sent for Ax and none for M and applying modus ponens we could deduce:

~deducible(sent,f)

which would not be an intended consequence of (vi).

The correct way to represent the information that sent has no models is:

~$\exists$M is-model(sent,M)
        or equivalently
$\forall$M ~is-model(sent,M)

4.3.5 The Importance of a Semantics

We have expended a lot of effort in this chapter on defining a *semantics* for all our logical expressions. That is, we have given recursive definitions of what sort of expressions are allowed and, with the aid of semantic trees, explained what they mean. This effort was not wasted. It was this we were relying on in the last section when we rejected a way of representing negation, which looked plausible, but could have stored up trouble for the future. It is this we will be using in chapter 16, when we *prove* that the theorem proving processes we design will work correctly.

This ability, to judge representations and processes, is not an idle luxury. The Artificial Intelligence literature abounds with plausible looking formalisms, without a proper semantics. As soon as you depart from the toy examples illustrated in the paper, it becomes impossible to decide how to represent information in the formalism or whether the processes described are reasonable or what these processes are actually doing.

Consider, for example, the *beta structures* of the Merlin system [Moore and Newell 73]. A beta structure is a way of describing concepts, e.g. in [Cunningham 78] the concept 'monkey' is described by:

monkey: (mammal (hands 4) tail intelligent).

which we are told 'means', mammal *further specified* by (hands 4), tail and intelligent. 'mammal' here looks like the *type* of any monkey. 'hands' looks like a function, which applied to a particular monkey will always return 4, e.g. hands(jacko)=4. 'tail' and 'intelligent', on the other hand, look like unary predicates, e.g. tail(jacko) & intelligent(jacko), meaning 'has a tail' and 'is intelligent', respectively.

Also in [Cunningham 78] 'ape' is described by

ape: (mammal (tail none) intelligent (hands 4)).

Here, however, tail has started to behave like a function, e.g. tail(washoe)=none, where none is an alias of 0.

A process of *matching* is described by which new descriptions can be generated from old ones. Matching can be achieved in several ways, in particular, if the beta structures, (heada furthera) and (headb furtherb), describing a and b have the same head word, i.e. heada≡headb, and the further specification of b is an extension of the further specification of a then b can be described by (a extras) where extras are the further specifications in furthera which are not in furtherb.

This process applies to monkey and ape above, since they both have the same head word, 'mammal' and ((tail none) intelligent (hands 4)) is an extension of ((hands 4) tail intelligent). The extra bit is (tail none), so ape can be described by

ape: (monkey (tail none))   (vii)

It is not clear whether (vii) is an intended consequence of matching. Its obvious interpretation, that apes are a type of monkey without tails, is strictly false in the real world, whereas the obvious interpretations of the original descriptions are not. However, it has a metaphorical interpretation that apes are rather like monkeys without tails, which is reasonable.

The fault of the Merlin system is that it is not clear what the intended interpretation of the formalisms are, and while we may be able to guess one when the concepts described are everyday objects, we may lose this ability as things get more complicated. Worst of all we cannot judge the correctness of processes like matching, because the meanings of the structures they manipulate are unclear. As long as the meaning is left unspecified by the authors of Merlin a user is free to choose any meaning whatever for his descriptions: and this may prove fatal if his choice violates some implicit assumptions behind the matching process, e.g. the kind of 'mistake' exhibited above may be much more serious for him.

There are times when the meaning of even the toy examples given in [Cunningham 78] elude me. I could understand

psum: (+ 3 para1 4)

as a term with dominating function +, where psum is presumably an acronym of 'partial sum'. However, one of the other Merlin processes returned the 'value' of this as

(7 para1)

and here I am at a loss. What is 7? If it is a constant then how can it be *further specified?* Can it be a function or a type?

Note that these last two beta structures overthrow all our assumptions about what kind of thing a beta structure can be and make us look again at the matching process above. Is it still correct? Can it withstand another assault like this? Only a semantics could provide an answer.

4.4 Summary

This chapter has covered a variety of topics.

- We considered higher order logic, involving variable functions, predicates, functionals, etc. We looked particularly at the typed lambda calculus.
- We gave sets of axioms for: equality, groups and natural numbers.
- We gave some hints on representing mathematical knowledge, including: whether to use functions or predicates; how to represent variadic functions and predicates; how not to represent negation and why semantics is important.

Further Reading Suggestion for Part I

We have now introduced all the mathematical logic necessary to have a formal notation for describing mathematical knowledge to a computer. For further reading in mathematical logic the reader is referred to [Mendelson 64].

Part II:
Uniform Proof Procedures

5
Formalizing the Notion of Proof

☐ This chapter defines 'proof' formally and introduces clausal form
and two rules of inference: resolution and paramodulation.
☐ Section 5.1 defines the resolution rule of inference in easy stages.
☐ Section 5.2 introduces Kowalski form.
☐ Section 5.3 defines the paramodulation rule of inference.

Now we have developed our formal notation for expressing mathematical
knowledge we can return to the question of how it can be manipulated in a
computer program. That is, we can continue our look at *procedures* for
'doing' Mathematics. In section 2.2.4 we described a decision procedure for
identifying correct propositional arguments. But in section 3.4 we saw that
there were practical difficulties in extending this procedure to arguments
involving the internal structure of propositions. So in this chapter we
develop an alternative technique: instead of trying to show that the
conclusion of an argument is a logical consequence of its hypothesis, we will
try to *prove* the conclusion from the hypothesis.

In order for a computer to prove theorems it is necessary that it should be
able to represent proofs internally. So we must extend our notation for
expressing mathematical knowledge to include proofs.

Definition 1: A proof is a sequence of formulae, each of which is either
an axiom or follows from earlier formulae by a rule of inference.

This definition captures the idea that to prove a theorem: we start with some
axioms; apply a rule of inference to derive a lemma from these axioms; add
the lemma to the original axioms; and continue recursively until we
arrive at the theorem we wanted. For instance,

1. $X=X$	(axiom)
2. $U=V \rightarrow (W=V \rightarrow U=W)$	(axiom)
3. $X=X \rightarrow (W=X \rightarrow X=W)$	(from 2. by substitution)
4. $W=X \rightarrow X=W$	(from 3. by modus ponens)

The skill of a mathematician consists in picking an appropriate rule of
inference and applying it in such a way that he arrives quickly at the theorem

he was trying to prove. This aspect is not covered by the definition, but it is the main theme of the rest of this book.

The theorem proving procedures described in this book follow the following pattern. The hypotheses, Hyp, of the theorem to be proved are added as temporary additional axioms to those, Ax, of the theory we are working in. A proof by contradiction is then sought, by adding the negation of the conclusion of the theorem, Conc, as an additional temporary axiom, and trying to prove the formula, f. If this succeeds then we have proved

$$Ax \,\&\, Hyp \,\&\, {\sim}Conc \rightarrow f$$

which is equivalent to

$$Ax \rightarrow (Hyp \rightarrow Conc)$$

Procedures based on this pattern are called *refutation systems*. We will see that refutation systems can direct the search for a proof, to some extent, by eliminating some of the more ridiculously inappropriate steps.

In this chapter we introduce two rules of inference, for deriving new theorems from old: *Resolution* and *Paramodulation*. Resolution is the Sheffer stroke of rules of inference: in a refutation system, it is the only rule of inference that is necessary in order to find proofs to all correct arguments. We say that resolution is *complete*. The proof of this, called the *completeness theorem*, is too technical for this introductory chapter, but is given in chapter 16. Paramodulation is introduced as an alternative to the axioms of equality. Both rules of inference apply only to First Order Logic, i.e. variable functions, variable predicates, variable functionals, etc are not allowed. In this chapter then, and until further notice, we restrict ourselves to First Order Logic.

Resolution and Paramodulation apply to only a restricted class of first order formulae, those in *clausal form*. This is not a significant restriction, since all mathematical theories and theorems can be translated into an equivalent clausal form. Again, the proof of this is too technical for an introductory chapter, and is relegated to chapter 15.

Clausal form is a normal form for logical formulae, in which some of the redundancy of the representation is eliminated. The use of a normal form for formulae, assists the computer to find proofs, by eliminating some duplication of effort. It helps avoid the situation where the computer develops two, apparently different, but essentially identical proofs, consisting of sequences of pairwise equivalent formulae. The equivalent formulae translate into the same normal form and the apparent difference between the proofs, disappears.

Definition 2: Clausal Form.
A set of formulae is in *clausal form* iff each formula is a *clause*. A clause is a disjunction of *literals*,
L_1 v...v L_n
where each literal is either a proposition or a negated proposition, e.g.

x=0 v x⩾0 v ~x=2

is a clause consisting of 3 literals, x=0, x>0 and ~x=2.

Note that clauses contain no quantifiers. Universal quantification is indicated by leaving variables free. Existential quantification is indicated by the introduction of new functions and constants called Skolem functions and Skolem constants.

5.1 The Resolution Rule

What is this resolution rule, about which we have heard so much? Let me start by explaining it in the simplest case, when no substitutions are required, and then gradually deal with more complex cases until we reach resolution in its full glory.

5.1.1 Stage 1 – Variable Free Resolution

Suppose we have two clauses, one containing the literal P and the other the literal ~P, where C' and C" are the respective remaining disjoined literals. The *resolution* rule enables us to deduce C' v C", i.e.

1. C' v P
2. C"v ~P

3. C' v C"

1 and 2 are called the *parents* of the *resolvant*, 3. P and ~P are called *complementary literals*.

Exercise 12: Check that 3 is a logical consequence of 1 and 2, using the techniques of section 3.4.

It may help you verify this rule in your own mind if you rewrite 1 as ~C' → P and 2 as P → C", deduce ~C' → C" and rewrite this to C' v C".

For any formula A, A v f is equivalent to A, by the semantic tree for v. Hence we adopt the convention that a disjunction of no literals is equivalent to f. In the definition of resolution above, either C' or C" or both may contain no literals. We say that they are *empty*. If both C' and C" are empty then we can derive the *empty clause*, denoted by f. This is the aim of a refutation system, and hence having C' and/or C" empty is a Good Thing.

Example

The two clauses:

$\sim$y=y v $\sim$x=y v y=x   and $\sim$y=x

contain the complementary literals, y=x and $\sim$y=x, where C' is $\sim$y=y v $\sim$x=y and C" is empty. Thus we can resolve the clauses to form:

$\sim$y=y v $\sim$x=y

5.1.2 Stage 2 – Binary Resolution

Now the parent clauses may not contain complementary literals, but it may be possible to apply substitutions to each of them so that the resulting clauses do, i.e. we may have clauses C' v P' and C" v $\sim$P" such that P'$\varphi'\equiv$ P"φ" for some φ' and φ", where $\equiv$ means 'is identical to'. In this case we can deduce (C' v C")$\varphi'\varphi$". It is usual to combine φ' and φ" into a single substitution, φ. Since a major objective of resolution is to minimize the amount of substitution going on, we will want φ to be the most general substitution such that P'$\varphi\equiv$P"φ. Fortunately, there is only one such, most general φ (up to permutations of variables) and there is a procedure for calculating it, given P' and P", called *unification.* φ is called the *most general unifier* of P' and P". Details of the unification procedure are too technical for this introductory chapter, and are given in chapter 17.

So the rule now is:

1. C' v P'
2. C" v $\sim$P"
---
3. (C' v C")φ

where φ is the most general unifier of P' and P".

This is called the *binary resolution rule.*

Example

The two clauses:

$\sim$U=V v $\sim$W=V v U=W   and $\sim$y=x

contain the literals, U=W and $\sim$y=x, respectively, which can be made complementary by the substitution {y/U, x/W}. Thus we can resolve them to form:

$\sim$y=V v $\sim$x=V

The newly derived clause should have its variables *standardized apart* before it is used as a parent of a new resolution, that is, its variables should be given new names, e.g.

$$\sim y = V' \text{ v } \sim x = V'$$

The reason for standarizing apart is to stop spurious coincidences preventing unification, e.g. $p(X)$ and $p(X+1)$ will not unify, whereas $p(X)$ and $p(Y+1)$ will.

5.1.3 Stage 3 – Full Resolution

One last wrinkle. It may be possible to eliminate, not just one, but several literals from the parent clause, at a stroke. Suppose the parent clauses are $C' \text{ v } P'_1 \text{ v}...\text{v } P'_m$ and $C'' \text{ v } \sim P''_1 \text{ v}...\text{v } \sim P''_n$ and φ is a most general unifier of $P'_1,...,P'_m, P''_1,...,P''_n$ then we may deduce $(C' \text{ v } C'')\varphi$ i.e.

1. $C' \text{ v } P'_1 \text{ v}...\text{v } P'_m$
2. $C'' \text{ v } \sim P''_1 \text{ v}...\text{v } \sim P''_n$

3. $(C' \text{ v } C'')\varphi$

where φ is the most general unifier of all P's and P"s.

This is called the *full resolution rule*.

Example

Consider the two clauses:

$$\sim U = V \text{ v } \sim W = V \text{ v } U = W \quad \text{and} \quad \sim y = X \text{ v } \sim Y = x$$

The first clause contains the literal $U = W$, and the second the literals $\sim y = X$ and $\sim Y = x$. These 3 propositions can be unified with the substitution, $\{y/U, x/W, x/X, y/Y\}$. Thus we can resolve the two clauses to form:

$$\sim y = V \text{ v } \sim x = V$$

Exercise 13: Apply the full resolution rule, in all possible ways, to the following pairs of clauses

1. $p \text{ v } q \text{ v } \sim r$ $s \text{ v } \sim p$
2. $p(X) \text{ v } q(f(X))$ $\sim r(Y) \text{ v } \sim p(f(Y))$
3. $p(X,a) \text{ v } p(f(a),Y) \text{ v } q(X,Y)$ $r(Z) \text{ v } \sim p(f(Z),Z)$

NB You can apply the rule to pair 3 in three different ways.

5.1.4 Factoring

In some Resolution systems, resolution is divided into two rules: binary resolution and *factoring*, i.e.

1. $C \vee P_1 \vee ... \vee P_n$

2. $(C \vee P_1)\varphi$

where φ is the most general unifier of all the Ps.

This is just a special case of the substitution rule.

Example
  Consider the clause:

$\sim y = X \vee \sim Y = x$

This contains the literals $\sim y = X$ and $\sim Y = x$, which can be unified with the substitution $\{x/X, y/Y\}$. Thus, with $\sim y = X$ as P_1, $\sim Y = x$ as P_2 and C empty, we can factor the clause to form:

$\sim y = x$

Everything we have to say about Resolution systems applies to both versions—the one rule or two rule systems.

5.2 Kowalski Form

Representing clauses as disjunctions of literals is convenient for theoretical discussions, but when we come to do actual resolution proofs (see chapter 6), a variant of clausal form due to Bob Kowalski and called *Kowalski form* will be more natural.

Definition 3: Kowalski Form.
  A clause,

$\sim P_1 \vee ... \vee \sim P_m \vee Q_1 \vee ... \vee Q_n$

where $P_1,..., P_m, Q_1,..., Q_n$ are propositions and m and n are ≥ 0 is in the, logically equivalent, *Kowalski form*, when it written as:

$P_1 \& ... \& P_m \rightarrow Q_1 \vee ... \vee Q_n$

That is, all the unnegated literals are collected in a disjunction of propositions on the right of an implication arrow, called the *consequent*, and all the negated literals are collected in a conjunction to the left of the arrow, called the *antecedent*.

Some authors draw the arrow pointing to the left with the propositions switched round, i.e.

$$Q_1 \text{ v}...\text{v } Q_n \leftarrow P_1 \&...\& P_m$$

We prefer the former notation since it preserves the format of the implication arrow.

The meaning of $\rightarrow$ must be extended to deal with the cases where either m or n is 0. When m is 0 we will interpret

$$\rightarrow Q_1 \text{ v}...\text{v } Q_n$$

as meaning

$$Q_1 \text{ v}...\text{v } Q_n$$

When n is 0 we will interpret

$$P_1 \&...\& P_m \rightarrow$$

as meaning

$$\sim\{P_1 \&...\& P_m\}$$

When both m and n are 0, we will interpret the empty clause, $\rightarrow$, as f, i.e. false.

Exercise 14: Go back to section 4.2 and check how many of the axioms are clauses in Kowalski form.

In this form the full resolution rule is

1. $H' \rightarrow C' \text{ v } P'_1 \text{ v}...\text{v } P'_m$
2. $H'' \& P''_1 \&...\& P''_n \rightarrow C''$

3. $(H' \& H'' \rightarrow C' \text{ v } C'')\varphi$

where φ is the most general unifier of the P's and P''s.

To see that this rule is true note that applying φ to 1 gives a formula equivalent to

$$H'\varphi \& \sim C'\varphi \rightarrow P$$

where $P \equiv P'_i\varphi \equiv P''_j\varphi$ Similarly, applying φ to 2 gives a formula equivalent to $P \rightarrow \sim H''\varphi \text{ v } C''\varphi$

From these we can deduce

$$H'\phi \ \& \sim C'\phi \rightarrow \ \sim H''\phi \ v \ C''\phi$$

which is equivalent to 3.

Example
  Consider the two clauses:
  $$U=V \ \& \ W=V \rightarrow U=W \quad \text{and} \quad y=X \ \& \ Y=x \rightarrow$$

The first clause contains the proposition $U=W$ in its consequent. and the second the propositions $y=X$ and $Y=x$ in its antecedent. These 3 propositions can be unified with the substitution, $\{y/U, x/W, x/X, y/Y\}$. Thus, with $U=V \ \& \ W=V$ as H' and C', H" and C" empty, we can resolve the two clauses to form:

$$y=V \ \& \ x=V \rightarrow$$

   An especially important class of clauses, both because they arise so often in practice and because simplified theoretical results apply to them are the *Horn clauses,* named after Alfred Horn, who originally isolated them. These are clauses with at most one unnegated literal, i.e those clauses in one of the four forms:

  Implication Clause
  $$P_1 \ \& ... \& \ P_n \rightarrow P$$

  Goal Clause
  $$P_1 \ \& ... \& \ P_n \rightarrow$$

  Assertion Clause
  $$\rightarrow P$$

  Empty Clause $\rightarrow$

  Exercise 15: Go back to section 4.2 and check how many of the axioms are Horn clauses.

5.3 The Paramodulation Rule

We have seen in section 4.2 that the equality axioms are a common ingredient of mathematical theories. Unfortunately, their inclusion in a set of clauses input to a Resolution theorem prover is a source of many difficulties, one of which is illustrated below.
   Suppose that we have derived

$$f(a)=b \rightarrow \text{ and } \rightarrow c=a$$

and wish to substitute c for a in $f(a)=b$ to form

f(c)=b →

Using the resolution rule this requires that the term, a, which is to be replaced must be isolated in an equation of its own, which can then be resolved with → c=a. This requires several unpacking steps: each one of which reduces the depth of the term, followed by a reorientation and then the main resolution, namely
- resolving with transitivity to derive f(a)=V & V=b →
- and resolving with substitution to derive a=V2 & f(V2)=b →.
- The term is now isolated, but needs reorienting using symmetry to V2=a & f(V2)=b →, before it can be resolved with → c=a.
- The resolution with → c=a can now take place producing f(c)=b →.

Thus a lot of rather unnatural looking steps are required to effect a rather simple substitution of equals for equals. One way round this problem is to make the substitution of equals for equals an inference rule of logic, i.e. to introduce a rule

1. C(T)
2. T=S
----
3. C(S)

where C(T) is a clause containing a term T and S is another term. This is the simplest case of a rule we will call *paramodulation*.

We will encounter a further problem with the equality axioms: their tendency to increase dramatically the branching rate of search trees. In chapter 7 we will consider this problem and whether paramodulation offers a solution to it.

The first wrinkle we will add to paramodulation is to allow only one occurrence of T to be replaced by S, instead of all of them. Several occurrences can then be replaced by repeated application. We will indicate that C contains a distinguished occurrence of T by using square brackets, i.e., C[T].

The second wrinkle is to recognize that the literal defining the substitution, T=S, is unlikely to appear in a clause on its own. It may have other disjuncts and these must be inherited by the derived clause.

The third wrinkle is to allow the substitution literal to be either way round, i.e. either T=S or S=T. The paramodulation rule now looks like

1. C[T]
2. T=S v D (or S=T v D)
----
3. C[S] v D

Finally we will let the two occurrences of T in 1. and 2. be unifiable rather than identical.

> 1. C[T']
> 2. T=S v D (or S=T v D)
> _____
> 3. (C[S] v D)θ

where θ is the most general unifier of T' and T.

The paramodulation rule can be used to replace nearly all the equality axioms, that is if we allow derivations using paramodulation as well as binary resolution and factoring then the equality axioms of symmetry, transitivity and substitution can be omitted. The only axiom we must retain is reflexivity, X=X. In addition we must add a family of new axioms, called the *functional reflexive axioms*, namely

$$f(x_1, \ldots, x_n) = f(x_1, \ldots, x_n) \tag{i}$$
for each n-ary function f

The functional reflexive axioms are something of an embarrassment to researchers in theorem proving: repeated attempts to prove the completeness of paramodulation + resolution without them have failed; but everyone's conviction is that they are really superfluous. The paramodulation completeness proof is complicated enough even with the functional reflexive axioms and we omit it (it is given in [Robinson and Wos 69]). The conviction that completeness can be proved without these extra axioms is called the *Paramodulation conjecture*. For an attempted proof of the result, now believed to be faulty, see [Richter 74].

5.4 Summary

We have defined a formal notion of proof as a sequence of formulae, starting with axioms and continuing with theorems derived from earlier members of the sequence by rules of inference. We have defined two such rules of inference: resolution and paramodulation. The first of these rules is sufficient to prove all correct arguments. The proof of this is given in chapter 16. The second is an alternative to the equality axioms.

To prove theorems by resolution, they and the axioms must be put into clausal form. Most of the axioms of section 4.2 are already in clausal form. A procedure for putting other formulae into clausal form is given in chapter 15.

Application of the resolution rule involves the unification of two or more literals, i.e. the calculation of the most general substitution which instantiates the two literals to a common instance. A procedure for making this calculation is given in chapter 17.

6
Searching for a Refutation

Now we have some rules of inference and some axioms, we can consider putting them together to generate some proofs. In this chapter we will consider the problems of getting a computer program to do this.

We will start by considering a simple resolution proof: the proof of the symmetry axiom for = from the reflexivity axiom and a slightly twisted version of the transitivity axiom (see section 4.2.1). The reflexivity and twisted transitivity axioms, put in Kowalski form are:

1. $\rightarrow X=X$
2. $U=V \ \& \ W=V \rightarrow U=W$

These are just as in section 4.2.1 except that the variables have been standardized apart and the second literal of the transitivity axiom has had its parameters switched around.

The conjecture we want to prove is the symmetry axiom $X=Y \rightarrow Y=X$. To prove a conjecture in a refutation system it must be: negated, put in clausal form , added to the axioms, and the empty clause derived. To ensure that the negation is done correctly, the conjecture must first have all its free variables bound by universal quantifiers, e.g.

$\forall X \ \forall Y \quad X=Y \rightarrow Y=X$

This process is called forming the closure. The conjecture can then be negated.

$$\sim\forall X\ \forall Y\quad X=Y\rightarrow Y=X$$

To put this formula in clausal form: the negation sign must be brought inside the universal quantifiers, changing them to existential quantifiers;

$$\exists X\,\exists Y\quad \sim\{X=Y\rightarrow Y=X\}$$

the existentially bound variables, X and Y, must be replaced with new constants, x and y, respectively;

$$\sim\{x=y\rightarrow y=x\}$$

the negation sign must be taken inside the implication arrow;

$$\sim\{\sim x=y\ v\ y=x\}$$
$$\sim\sim x=y\ \&\ \sim y=x$$
$$x=y\ \&\ \sim y=x$$

and the formula separated into two clauses in Kowalski form.

3. $\rightarrow x=y$ and 4. $y=x\rightarrow$

A general procedure for transforming a formula into clausal form is given in chapter 15 together with a proof that a formula has a model iff its clausal form has a model. These give a technical justification of the above steps.

An intuitive justification is as follows. In order to prove a conjecture of the form $P\rightarrow Q$, we assert $\rightarrow P$ as a new axiom and set up $Q\rightarrow$ as a goal to be proved. Free variables in conjectures, like X and Y, are regarded as 'typical' objects, for which, if P is true then Q is true. In a refutation system, such 'typical' objects cannot be represented by free variables, or else there is a danger that a faulty 'proof' will be found. For instance, if clause 4 were

4'. $Y=X\rightarrow$

then the empty clause could be derived by resolution with either

1. $\rightarrow X=X$ or even $\rightarrow 0=0$

without reference to the other axioms. To prevent this kind of faulty 'proof' we must represent 'typical' objects by constants, but not existing constants, like 0 or 3. We must create new constants, x and y, about which nothing is already known. x and y are called *Skolem constants* after Thoralf Skolem who first invented this device. We will adopt the convention of using the lower case version of an existential variable when converting it into a Skolem constant.

We are now in a position to derive a proof of the conjecture. We have four axioms: 1, 2, 3 and 4.   4 and 2 can be resolved to form

   5. y=V' & x=V' →

and 3 can be resolved with this to form

   6. y=y →

Finally, 1 can be resolved with 6 to form

   7. →

the empty clause.

   Notice how this refutation takes the result of each resolution and resolves this in turn with one of the original clauses. We can represent this situation with the path in figure 6-1 where the nodes are labelled with the goal clause and the subsequently derived clauses, and the arcs are labelled with the numbers of the original clauses.

6.1 Following Your Nose

This kind of refutation is very natural. It models, in a simple way, the 'mental set' of a real mathematician who takes the original conclusion of a conjecture and, working backwards, worries away at it with axioms and hypothesis. It is called *Linear, Input Resolution*. *Linear* because we keep using the last resolvent as the parent of the next resolution and so can represent the refutation as a linear path: we do not resolve 4 and 2 and then,

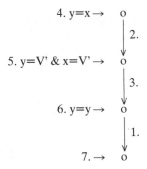

Figure 6-1: The Refutation as a Path

say 1 and 2. *Input* because we always resolve the last resolvent with one of the original, input clauses, not with a derived clause, e.g. having derived 6 we may resolve it with 1, 2, 3 or 4, but not with 5.

Linear, Input Resolution is an example of a *refinement* of full resolution.

Since it excludes various possibilities it is inherently more efficient than full resolution, i.e. it involves less search. It is complete for sets of Horn clauses (like those in our example), i.e. it will find a refutation if there is one. This follows from the Horn clause completeness of Lush Resolution below. For non-Horn clauses we must relax the Input constraint, but we can keep the Linear one. A non-Input resolution, e.g. 6 with 5, is called *ancestor resolution*, because 6 resolves with one of its ancestors higher up the path.

6.2 Representing Choice

Of course, the path in figure 6-1 is not the only possible one. Figure 6-2 is another one (which succeeds)

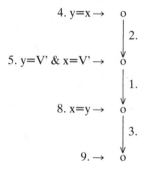

Figure 6-2: Another Refutation Path

and figure 6-3 another (which represents a loop).

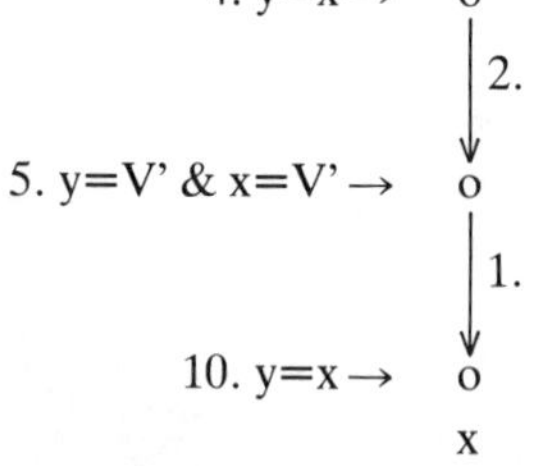

$$4.\ y=x \rightarrow \quad o$$
$$\Big|\ 2.$$
$$5.\ y=V' \ \&\ x=V' \rightarrow \quad o$$
$$\Big|\ 1.$$
$$10.\ y=x \rightarrow \quad o$$
$$x$$

Figure 6-3: A Loop

We can fit all these paths together to make a new kind of tree, a *search* tree.

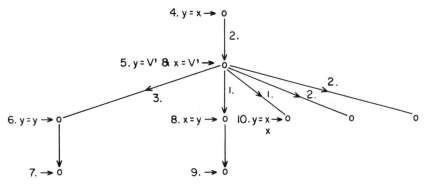

Figure 6-4: A Search Tree

Once again trees come to our aid as a descriptive device. This time to describe the process of searching for a refutation.

Exercise 16: Represent the applications of clause 2 to clause 5 by labelling the two unlabelled nodes in figure 6-4.

6.3 AND Choices and OR Choices

Consider the choices represented by the five way branching from clause 5 in figure 6-4. There are really two kinds of choice packed together here:

- the choice of which literal (or in general, which group of literals) to resolve away;
- the choice of which clause to resolve the literal away with.

   If the first literal is chosen then it can be resolved away in two ways: with 1 to yield

$x=y \rightarrow$

or with 2 to yield

   $y=V" \text{ \& } V'=V" \text{ \& } x=V' \rightarrow$

If the second literal is chosen then it can be resolved away in three ways: with 1, 2 or 3.

   These two kinds of choice are essentially different.

- The choice of literal is only a question of order: all literals must eventually be resolved away, we may only choose what order to do this in. We call such a choice an *AND* choice, because we must first choose one clause AND then another.
- The choice of which clause to resolve with is a genuine choice. If this branch of the tree leads to the empty clause then the other branch does not need to be developed. We call such a choice an OR choice,

because we may choose either one clause OR another.

AND choices are innocuous. It does not affect the completeness of Linear Input Resolution in what order the literals of a clause are resolved away. So we can pick an order, either at random or exercising extreme cunning, and stick to it. This means that branch points representing AND choices can be pruned from our search tree. The resulting tree for our running example is given in figure 6-5. The literal selected to be resolved away next is underlined.

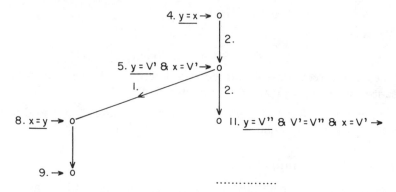

Figure 6-5: A Lush Resolution Search Tree

OR choices are not innocuous. They are a (the?) major source of difficulty in automatic theorem proving. We will address this difficulty in section 6.7.

Thus if our original clauses are all Horn clauses we may apply Linear Input Resolution with the following additional restriction; that a literal be selected from the current clause and that any resolutions between the current clause and the input clauses must involve that selected literal. The resulting Resolution restriction is called *Lush Resolution* [Hill 74]. Lush stands for *L*inear Res*olution* with *u*nrestricted *s*election function for *H*orn clauses. In the case of non-Horn clauses the choice of selected literal is restricted to be one of the most recently introduced. This restriction is called *SL Resolution* [Kowalski and Kuehner 71]. SL stands for *S*elected *L*iteral.

6.4 Preventing Looping

In figure 6-3 we give an example of a loop: the derivation of a clause identical to one derived previously. There is no point continuing to develop this clause, since it can only lead to clauses which could already have been derived from its predecessor: we will merely be duplicating work. We will certainly want to build a loop checker into our theorem prover, i.e. a piece of program which detects loops and prevents them from being developed further.

In fact, we can go further than this. Suppose the clauses

$$y=x \rightarrow \qquad\qquad\qquad\qquad\qquad\qquad (i)$$

and

$$Y=x \rightarrow \qquad\qquad\qquad\qquad\qquad\qquad (ii)$$

have been derived. Now since (i) is an instance of (ii) anything derivable from (i) can also be derived from (ii). Thus (i) is a kind of loop and further derivations from it should be blocked. (i) is said to be *subsumed* by (ii) and the resolution restriction of blocking further development of subsumed clauses is called *subsumption checking*.

The clauses

$$y=x \ \& \ a>b \rightarrow \qquad\qquad\qquad\qquad\qquad (iii)$$

and

$$y=x \rightarrow \qquad\qquad\qquad\qquad\qquad\qquad (iv)$$

exhibit another kind of subsumption relationship. Any derivation of the empty clause from (iii) can be pruned into a derivation of the empty clause from (iv) by deleting occurrences of $a \geqslant b$ and those resolutions required to resolve it and its descendants away. Thus (iii) cannot produce any refutations that cannot be produced better from (iv) and further derivations from it can be blocked without loosing completeness.

Definition 1: In general, a clause, C, subsumes a clause, D, if there is a substitution Θ such that

$$D \longleftrightarrow C\Theta \ v \ E$$

For instance,

$$\sim y=x \longleftrightarrow \sim(Y=x)\{y/Y\} \ v \ f$$

$$\sim y=x \ v \sim a>b \longleftrightarrow \sim(y=x)\{\} \ v \sim a>b$$

Subsumption checking preserves the completeness of most Resolution restrictions provided care is taken when the subsuming and subsumed clauses are identical. In this case the second clause derived must be the one whose further development is blocked or the looping may continue.

Otherwise, it does not matter whether the subsumed or subsuming clause is derived first.

Because we have chosen to represent the process of searching for a refutation with search *trees*, a clause which can be derived in more than one way is represented by more than one node. If we decide to identify nodes in

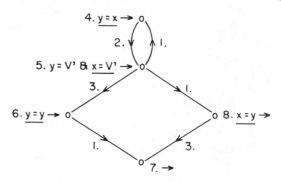

Figure 6-6: A Search Graph

the tree labelled by the same clause, then the resulting structure is called a *search graph.** Figure 6-6 is a graphical representation of the tree in figure 6-4.

When a loop check is built into the theorem proving program the graph representation of the search is particularly appropriate.

6.5 Choosing Where to Start

In all our example refutation paths we started with 4 as the *top clause*, so this is the root of the tree in figure 6-4. Starting with 4 seemed like a good idea. Since 4 was inherited from the conjecture, it seemed bound to be involved in any refutation. Using it as the top clause guaranteed that we didn't try to make a refutation involving just 1, 2 and 3. In fact, all four clauses are bound to be involved, so it would have been as good to start with any one of them.

{1,2,3,4} form a *minimally unsatisfiable* set of clauses. That is, they are unsatisfiable, but dropping any one of them will result in a satisfiable set. It is not possible to derive the empty clause from a satisfiable set of clauses, so 1, 2, 3 and 4 must all be involved in any refutation.

Suppose that the theorem prover were provided with an additional axiom, e.g.

*Contrary to common expectations the branches of a wooden tree can be made to rejoin by careful grafting. Maybe this is the origin of the term graph.

5. $\rightarrow$ X+1 > X

It would not have been a good idea to use 5 as top clause, because no other clause will resolve with it, and so the empty clause will not be derived. Put in another way, 5 is not a member of a minimally unsatisfiable subset of {1,2,3,4,5}. The top clause should always be chosen from a minimally unsatisfiable set. If a minimally unsatisfiable set cannot be detected then it is necessary to try each clause in turn as top clause. Only if all these refutation attempts fail can we be sure that the conjecture is not a theorem.

If a minimally unsatisfiable set can be detected then we only need try one of its members as top clause. If this refutation attempt fails then the conjecture is not a theorem. This is called the *set of support* strategy.

The usual trick for choosing the top clause, is the one we used, of choosing a clause inherited from the conjecture, e.g. 3 or 4. However, such a clause is not guaranteed to belong to a minimally unsatisfiable set, because the conjecture may not be theorem, or the clause we choose may not play a role in any proof. A trick like this, which is not *guaranteed* to succeed, but probably will, is called a *heuristic*.

> *Exercise 17:* Draw a Lush Resolution search tree, of at least size 6, with clause 3 as top clause.

When a goal clause, like 4, is chosen as top clause, the resulting refutation is called a *backwards proof* and the search process a *backwards search*, because they start from the conclusion of the theorem and work backwards to the axioms and hypothesis of the theorem.

When an assertion, inherited either from the theorem, like 3, or an axiom, like 1, is used as the top clause the refutation and search process are called a *forwards proof* and a *forwards search*, respectively.

6.6 Non-Horn Clauses, Case Analysis and Ancestor Resolution

To demonstrate the incompleteness of Lush Resolution when there are non-Horn clauses, consider the following example.

1. $\rightarrow$ natural(n) v non-neg(n)
2. natural(n) $\rightarrow$ non-neg(n)
3. non-neg(n) $\rightarrow$ natural(n)
4. natural(n) & non-neg(n) $\rightarrow$

natural(n) means n is a natural number. non-neg(n) means n is a non-negative number. The clauses can be interpreted as a request to show that n is both a natural number and a non-negative number (clause 4) under the assumptions that: it is either one or the other (clause 1); if it is natural then it is non-negative (clause 2); and if it is non-negative then it is natural (clause 3). Clearly, this should be possible.

However, clause 1 is a non-Horn clause. Let us choose the goal clause, 4, as top clause and consider the partial refutation in figure 6-7.

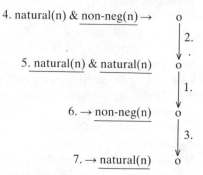

4. natural(n) & non-neg(n) →

5. natural(n) & natural(n)

6. → non-neg(n)

7. → natural(n)

Figure 6-7: A Non-Horn Clause Partial Proof

Clause 7 will only resolve with input clauses 2 or 4. Resolution with 2 will result in → non-neg(n), which is a loop. Resolution with 4 will result in non-neg(n) →, which is a symmetrical situation to the one we now face. Similar situations occur on all the other branches and so the empty clause cannot be derived.

Exercise 18: Develop the complete, Lush Resolution, search tree for clauses 1 to 4, using 4 as the top clause. Check that the tree is finite, but does not contain the empty clause.

The partial refutation in figure 6-7 has several interesting features.

- Note the use of full resolution to resolve away two occurrences of natural(n) simultaneously.
- Note also that clause 6 and 7 are not goal clauses, i.e. they have non-empty consequents. If the top clause is a goal clause and the input clauses are Horn clauses then this situation cannot arise. It has happened here because of the resolution with the non-Horn clause 1.

One way to deal with the incompleteness of Lush Resolution is to allow *case analysis* by splitting clause 1 into two clauses,

  1a. → natural(n)
  1b. → non-neg(n)

and seeking two Lush Resolution refutations, one from clauses 1a, 2, 3 and 4, and one from clauses 1b, 2, 3 and 4. The resulting refutations are shown in figure 6-8.

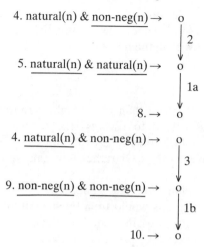

4. natural(n) & non-neg(n) →   o

5. natural(n) & natural(n) →   o

8. →   o

4. natural(n) & non-neg(n) →   o

9. non-neg(n) & non-neg(n) →   o

10. →   o

Figure 6-8: Proof by Case Analysis

An alternative technique is to relax the input restriction to allow ancestor resolution, i.e. the resolution of the current clause with one of its own ancestors. This allows us to resolve clause 7 in figure 6-7 with its ancestor, clause 5, to derive the empty clause. The result is exhibited in figure 6-9.

One way to think of this refutation is as a combination of the two case analysis refutations of figure 6-8. The steps of the two case analysis refutations are input resolutions with 2 and 1a and with 3 and 1b. Inverting the second pair, and merging 1a and 1b to give 1, gives a sequence of resolutions with 2, 1 and 3. These are the steps of the ancestor resolution

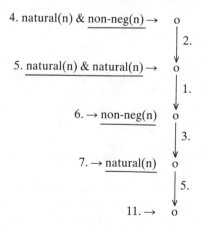

4. natural(n) & non-neg(n) →   o

5. natural(n) & natural(n) →   o

6. → non-neg(n)   o

7. → natural(n)   o

11. →   o

Figure 6-9: An Ancestor Resolution Refutation

refutation. The use of ancestor resolution in the last step can be read as the ability to assume ~natural(n) in the second case, since the case where natural(n) has already been dealt with in the first case.

6.7 How to Make OR Choices

In a search tree, every node with more than one daughter represents a choice. Following one path may lead to success (the empty clause). Following another may lead to failure (only loops and non-empty clauses without daughters). How can we choose? Or rather how can we instruct our computer programs to choose?

We could let them use random number generators – a sort of computational coin toss. The programs would then be simulating one of the 'mysterious' processes mentioned in the introduction – luck! (Certainly the easiest of the 'mysterious' processes to simulate.) Or we could try to find heuristics to guide the search, and simulate instead the remaining 'mysterious' processes, of experience and intuition. Finding such heuristics is certainly the hardest problem in automatic theorem proving!

6.7.1 Depth First Search

Let us start by following up the ideas of 'mental set' discussed in section 6.1. We claimed that a natural way to search for a refutation was to take the last clause derived and employ it as the parent of the next resolution. This suggests a search strategy in which a path is pursued until this development is blocked in some way: either by success, the derivation of the empty clause: or by failure, a loop having been detected or no further resolution being possible. In the case of failure we will then want to back-up and try some other path. If we stick to our desire to model mental set we will want to retreat as little as possible – only as far as the last choice point – then plunge on down again.

The search strategy just described is called depth first search. It is illustrated in figure 6-10. Each node is labelled with a letter indicating the order in which it was developed. (The actual clauses are irrelevant here and are omitted.) A cross represents some kind of failure, causing back-up.

Back-up from node F passed through nodes E and C, back to B, indicating that no further choices were available at E and C.

Depth first search leaves mostly unanswered the central question of automatic theorem proving: how the choices are to be made. It gives no grounds for the choice of developing C before G, but once this is made it forces the choice of F before H. The unforced choices may be made

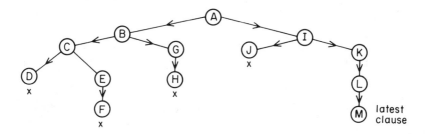

Figure 6-10: Depth First Search

arbitrarily or with extreme cunning: the search will still be depth first provided two criteria are met.

– The current path is pursued until further progress is blocked.

– Back-up is always to the latest choice point.

6.7.2 Breadth First Search

Not only does depth first search fail to settle the central question, it fails to answer two simpler questions.

– *Complete Search*
How can the tree be searched so that every node is eventually reached? In depth first search we may pursue an infinite branch indefinitely and never back-up to the rest of the tree. Thus we may choose a complete inference system (e.g. Linear Resolution), but still invalidate our guarantee of being able to prove all theorems, if we then choose an *incomplete* search strategy like depth first search.

– *Shortest Proofs*
How can the tree be searched so that the shortest refutation is found first? In depth first search we may find a solution on one branch, but a shorter refutation may be lurking, undiscovered, on another, unsearched branch.

One answer to both these questions is provided by *breadth first search.* In breadth first search all the nodes at depth 1 are developed first, then all the nodes at depth 2 and so on (see the illustration in figure 6-11 in which the labels on the nodes again illustrate the order in which they were developed).

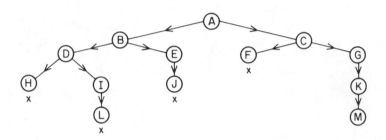

Figure 6-11: Breadth First Search

A breadth first search strategy develops the search tree in stages: at stage n all nodes of depth n are developed. Since there are only a finite number of nodes with depth n, stage n will terminate eventually, and stage n+1 will commence. Hence,

- All nodes will eventually be developed. A node of depth n, will be developed during stage n.
- The first refutation to be found will be shorter than any other. If the length of the shortest refutation is n then it will be found at stage n before any other refutation is found.

These arguments confirm that breadth first search answers the questions about complete search and shortest refutations. It also provides a partial answer to the central question – each node is developed during stage n – the only remaining choice is the order of development of nodes within each stage.

Thus breadth first search provides answers to all our questions, but not very intelligent answers. The strategy has no psychological plausibility. People never search for refutations like this. Besides the abandonment of 'mental set' – the flipping about from one path to another – this method has huge memory overheads. At stage n, all the nodes of depth n must be remembered. The number of such nodes usually grows exponentially! For instance, if each node has approximately 10 daughters there will be of the order of a million nodes at depth 6. This is quite a problem for computers too. They soon get bogged down in the memory overhead.

6.7.3 Heuristic Search

Depth first and breadth first search are both *uninformed* search strategies: that is, they both search the search tree without regard to the particular problem being solved. In this section we consider how knowledge about the problem can be used to influence the direction of the search.

When a human is searching for a refutation he sometimes has a feeling he is on the right track and sometimes that he is hopelessly lost. Some steps seem to take him closer to the goal and some further from it. How can we capture these feelings and use them to guide the search?

One simple and crude technique is to associate a number with each node of the search tree, called its *score*, where the smaller the score the better the node. Whenever we have a choice, of which node to search next, we choose the node with the smallest score. This search strategy is called *heuristic search*. It is illustrated in figure 6-12. The score of each node is written beside it and the letters, as usual, indicate the order in which the nodes are developed. Lower case o, indicates a node which is *open*, i.e. the undeveloped daughter of a developed node. Note that, at each stage, the next node developed is always the open node with the lowest score. In programming heuristic search, open nodes are usually stored on a list, called an *agenda*, in increasing order of score. The node to be developed next is, thus, the first node on the agenda.

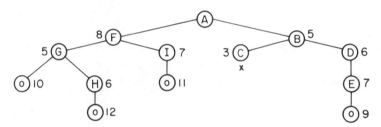

Exercise 19: Which node should be developed next in figure 6-12?

Where do the numerical scores come from? The score is assigned to a node by applying to it an *evaluation function*. The evaluation function is heuristic. It may be general purpose or special purpose, i.e. it may be useful for guiding the search for the refutation of any conjecture from any set of axioms or it may exploit special properties of the domain it is used in.

Length of clause is an example of a general purpose evaluation function. The length of a clause is the number of literals it contains. The idea of this evaluation function is that the empty clause has length 0 and that the shorter the clause the fewer literals have to be resolved away to derive the empty clause – thus the shorter the better. Suppose the current clause were

y=V' & x=V' →

and we can resolve the first literal with the reflexivity axiom to produce

$$x=y \rightarrow \qquad \qquad \text{(v)}$$

or with the twisted transitive axiom to produce

$$y=V'' \ \& \ V'=V'' \ \& \ x=V' \rightarrow \qquad \qquad \text{(vi)}$$

Now clause (vi) has length 3, whereas clause (v) only has length 1 and hence will be developed first by heuristic search. Conjectures can be found in which this would be the correct choice and others for which it would be the wrong choice, but on the whole it is sensible practice.

A more elaborate evaluation function could be defined by assigning a subscore to each literal based on its complexity and then summing these subscores. The depth of a literal considered as an expression tree is one simple measure of its complexity. This is sometimes called the *depth of function nesting*. The depth of $x=y$ is 1 and the depth of $x+1=2.y-1$ is 3, so that the score of

$$x=y \ \& \ x+1=2.y-1 \rightarrow$$

is 4.

We could get more elaborate still, by assigning different subscores to different functions and predicates, e.g. $=$ might rate 1, $+$ and $-$ might rate 1 and . might rate 2; expressing a conviction that literals containing . are hardest to resolve away and those containing only $+$ are easier, etc. Such an evaluation function is special purpose because it relies, in part, on heuristics specific to the domain of arithmetic.

To find out more about heuristic search; including what conditions must be placed on the evaluation function to guarantee that shortest refutations are always found first; see [Nilsson 80].

6.8 Summary

We have been considering how to search for the refutation of a theorem by applying rules of inference to formulae in different ways. Such a search process can be described with the aid of search trees and search graphs.

In particular, we have looked at refining resolution by only considering refutations which obey the linear, input, selected literal or subsumption restrictions. In the case of non-Horn clauses we saw that the input restriction was too strong and that ancestor resolution had to be allowed. We saw that an additional rule of inference, paramodulation, made refutations involving equality shorter and more natural.

We also looked at various search strategies including: backwards versus forwards search; and depth first, breadth first and heuristic search.

Further Reading Suggestions for Chapters 5 and 6

The last two chapters have only scratched the surface of the immense amount of work on uniform proof procedures. [Chang and Lee 73] gives a more comprehensive survey. [Nilsson 80] is a general introduction to Artificial Intelligence problem solving techniques.

7

Criticisms of Uniform Proof Procedures

☐ This chapter reviews and evaluates the contribution of resolution type proof procedures to mathematical reasoning.

☐ Section 7.1 lists the ways in which mathematical logic has contributed to mathematical reasoning.

☐ Section 7.2 uses a group theory example to illustrate the combinatorial explosion.

☐ Section 7.3 describes various uniform techniques for avoiding or overcoming the combinatorial explosion, including a discussion of whether additional rules of inference, e.g. paramodulation, reduce the amount of search (7.3.1), and a discussion of ways of disguising how bad things are (7.3.2).

☐ Section 7.4 begins a discussion of non-uniform guidance techniques, starting with an analysis of a human proof of the group theory example.

☐ Section 7.5 formalizes the knowledge used in the human proof.

☐ Section 7.6 generalizes the last two sections into a new methodology for mathematical reasoning.

The previous chapters have described some of the work done in 'Automatic Theorem Proving' during the 1960s, especially the resolution and paramodulation rules of inference and the formalization of mathematical knowledge using the Predicate Calculus. The time has now come to take stock: to consider how effective these techniques are in enabling us to build artificial mathematicians; and to present some of the criticisms of such 'uniform proof techniques' levelled by other workers in Artificial Intelligence.

7.1 The Contribution of Logic

Automatic Theorem Proving was based on the solid foundation of earlier work in Mathematical Logic. This is not surprising: both fields had similar aims. Mathematical Logic was concerned to justify mathematical activity, that is to analyse and explain the existing proofs of Mathematics; relating fields together, discovering and correcting inconsistencies and omissions etc. Automatic Theorem Proving was concerned to explain how proofs could be discovered in the first place. Though similar, these aims are different, and we cannot expect that Logic should provide all the answers, or even that it should suggest the right approach, to the problem of explaining proof discovery.

Let us start by listing what I take to be the major contributions of Mathematical Logic to Automatic Theorem Proving.

1. The notation of Logic (see chapters 2, 3 and 4) gives us a precise language, suitable for describing mathematical knowledge to a computer.
2. The semantics of this language. This enables us to relate the notation to actual pieces of mathematics and to eliminate faulty representations (see especially section 4.3).
3. The definition of a proof as a sequence of applications of rules of inference to axioms or previously proved theorems. This enables us to be precise about what we are hoping to discover. Before the advent of Logic a proof was any convincing argument.
4. Hilbert's claim that any mathematical activity can be regarded as finding proofs in some formal system. For instance, equation solving can be regarded as theorem proving where our main interest is in the substitutions for variables on route. This gives theorem proving a central role in mathematical reasoning.
5. The standard axiomatizations of many mathematical theories. In section 4.2 we gave the axiomatizations of equality, groups and arithmetic and used some of these axioms to demonstrate the workings of the resolution and paramodulation rules.

We will see that the value of these contributions diminishes as we move down the list. Numbers 4 and 5 are especially suspect and two major themes of the chapters ahead will be: the investigation of areas of mathematical reasoning other than theorem proving; and the provision of non-standard axiomatic theories.

However, they have provided a useful first approximation to a solution. They have enabled us to design theorem proving systems and prove elementary theorems artificially. To improve on this and move on to a second approximation we must turn our attention to an actual resolution proof attempt and see in what ways it goes wrong.

7.2 A Resolution Proof and the Combinatorial Explosion

Consider the theorem in group theory that

Every group of exponent 2 is abelian.

This problem can be formalised in a way suitable for solution by a Resolution theorem prover by taking the clausal form of

    (a) the axioms of equality,

    (b) the remaining axioms of group theory,

    (c) the axiom, $XoX = e$,

    (d) and the negation of the commutativity axiom, $AoB = BoA$. (This is the goal clause, $aob = boa \rightarrow$.)

and trying to derive the empty clause from them using the resolution rule.

The complete set of clauses is:

    1. $\rightarrow X1 = X1$.

    2. $X2 = Y2 \rightarrow Y2 = X2$.

    3. $X3 = Y3 \ \& \ Y3 = Z3 \rightarrow X3 = Z3$.

    4. $X4 = Y4 \ \& \ Z4 = W4 \rightarrow X4 \ o \ Z4 = Y4 \ o \ W4$.

    5 $X5 = Y5 \rightarrow i(X5) = i(Y5)$.

    6. $\rightarrow X6 = e \ o \ X6$

    7. $\rightarrow X7 = X7 \ o \ e$.

    8. $\rightarrow e = X8 \ o \ i(X8)$

    9. $\rightarrow e = i(X9) \ o \ X9$.

    10. $\rightarrow X10 \ o \ (Y10 \ o \ Z10) = (X10 \ o \ Y10) \ o \ Z10$.

    11. $\rightarrow X \ o \ X = e$.

    12. $a \ o \ b = b \ o \ a \rightarrow$.

The empty clause can be derived from these clauses after 42 resolutions starting with number 12 as the top clause. In figure 7-1 we give the very beginning of the search tree generated by Lush Resolution.

From this it can be seen that the search tree rapidly grows as the depth increases. In fact, the average number of daughters per parent, or the *branching rate*, is 3. At this stage only the equality axioms are applicable. Later on some of the group theory axioms get a look in and the branching rate goes up.

Even with this, rather conservative, estimate of a branching rate of 3, there will be 3^{42} nodes at depth 42, where the first empty clause can be found. This is a very big number! It is of the order of 10^{21}. That number of clauses could not be stored in any one of todays digital computers, so a breadth first search strategy is obviously out of the question. A depth first or a heuristic search is feasible, but is unlikely to find the proof unless it receives very cunning guidance. There are too many losing choices available. This rapid growth of the size of the search tree with the depth is called the *combinatorial explosion*. It is a stick with which to beat uniform proof procedures.

This particular theorem has a special place in my heart. When I first got the opportunity to feed theorems to an automatic theorem prover this was the theorem I fed. It seemed about the right level of difficulty. Not trivial, but easy enough to be the sort of theorem to give to beginning students of group theory. Imagine my surprise when the SL Resolution theorem prover I was using had found no proof after 3/4 hour (I turned it off to put it out of its misery). I printed out the partial search tree to see what had gone wrong. All the steps were perfectly legal, but that was about all you could say for them! It was clear that without guidance the theorem prover would do the most amazingly silly things.

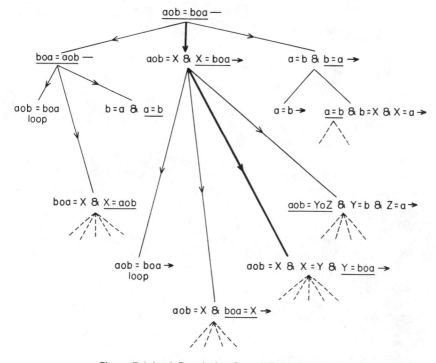

Figure 7-1: Lush Resolution Search Tree to Depth 2

My experience is pretty standard. An unguided theorem prover can prove trivial theorems by shear brute strength, but collapses under the weight of anything the least bit difficult. It is hard to imagine, in advance, what silly things it might try to do. We humans tend to be so prejudiced in favour of the proof we know is in there, that we overlook most of the blind alleys.

Let us look at some of the silly things which are going on even in the little bit of search tree in figure 7-1.

    – The path marked in heavy type consists of repeated applications of the transitivity axiom. The effect of these is to break equations into two, inserting a new variable in the openings thus created. We saw in section 5.3 that such a move is occasionally essential. However, repeated applications to no purpose are silly. They create a situation a little like looping, but which, unfortunately, cannot be picked up by a conventional subsumption check.

– The application of clause 4 to the top clause creates the goals a=b and b=a. These are certainly unprovable, since a and b are arbitrary, and hence unequal, constants. It is silly to pursue this branch any longer. (In chapter 10 we will describe a mechanism for automatically pruning such branches.)

7.3 Attempts to Guide Search

Clearly, a theorem prover which is going to tackle non-trivial theorems needs some guidance. A variety of techniques have been explored for providing such guidance. Let us look at some.

The most common way is to introduce some completeness preserving restriction to basic resolution. We have already seen some of these: the Linear and Input restrictions and the selection of literals; the use of set of support and the deletion of subsumed clauses. In fact we employed all of these in the search tree of figure 7-1 and still ran into trouble.

Another technique is to expand the inference system in the hope of finding shorter proofs or of decreasing the branching rate by being able to delete axioms. The introduction of the paramodulation rule was such an expansion. Let us see what effect its introduction has on our group theory example.

7.3.1 Paramodulation

The use of paramodulation telescopes our 42 step resolution proof into a 10 step proof. This is because, as shown in section 5.3, several resolution steps are required to make an equality substitution, and these steps collapse into one paramodulation step. The actual paramodulation plus resolution proof is given in figure 7-2.

A reduction from 42 steps to 10 steps is very good news and encourages us to think that the the automatic proof of the theorem may now be within the grasp of an automatic theorem prover.

The bad news is that the branching rate is also dramatically increased from about 3 to about 12. So we have gone from a deep thin tree to a shallow bushy tree. Since 12^{10} is less than 3^{42}, this seems like it might be an improvement, however the branching rates of 3 and 12 were only very crude estimates so we cannot be sure without a more careful analysis and we

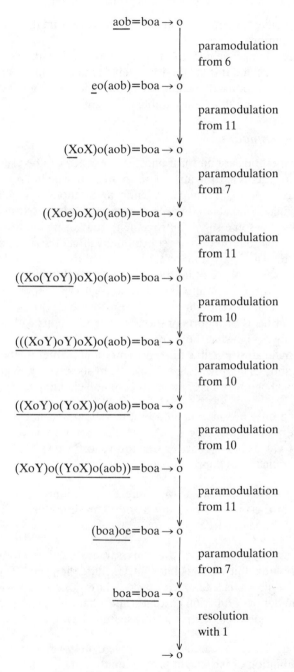

Figure 7-2: A Paramodulation and Resolution Proof

certainly cannot deduce anything about the general case from this. And 12^{10} is still a very big number.

There is a marked improvement in the naturalness and readability of the paramodulation proof and this *is* a positive step. It makes it much easier to bring to bear our intuitions about how the proof should be guided if the theorem prover is making steps comparable to our own.

7.3.2 Cheating Techniques

All the above methods are uniform ones: that is, they can be used when trying to prove any theorem from any set of axioms. They do not rely on any domain dependant proof techniques. Uniform methods are generally regarded as having failed. New inference systems, either restrictions or extensions, may offer some slight improvement, may bring another trivial theorem within reach, but they do not substantially affect the situation.

Nevertheless there is always someone, who after enormous effort locating and stamping out some source of inefficiency, believes that his new system offers a panacea. In the literature one may find glowing reports of such innovations. The inadequacies of such systems are easy to overlook or (horrors!) to cover up. The reports of successful experiments with the new system may not always be what they seem!

To forearm you against possible misrepresentations, either deliberate or unintentional, (and, as an unwitting side effect, to show you how to cheat yourself) we record below some of the most common cheating techniques.

- Displaying only proofs obtained by the system and not the search tree generated in the process. As we have seen above, it is often possible to trade a deep narrow tree for a short wide tree, but this is not automatically an improvement.

- Feeding to the theorem prover only those axioms known to be required in the proof. Irrelevant axioms can dramatically increase the size of the search tree.

- Generating only that part of the search tree which lies within some arbitrary limits, but which is known to include the proof. Typically, the sought proof is examined to see: how long it is; what the maximum depth of function nesting is and what the maximum length of clauses is: and then the tree is only searched within those limits.

- Of course, if such techniques are applied uniformly and without pre-examination of the sought proof then they are legitimate although they will always lose completeness). One could imagine, for instance, such a technique being proposed as a psychological theory of the memory limitations of human mathematicians.

– The program implementing the new system may be rejigged for each new proof by a variety of other techniques, e.g. changing the criteria by which literals are selected and clauses chosen.

We would not want to suggest that these techniques are peculiar to Automatic Theorem Proving. Their analogues can be found in all areas of Artificial Intelligence and, one gathers, in most other branches of Science.

So if uniform methods have failed, what are we to do?

7.4 Analysing Human Proofs

One way forward is to turn our attention to the way in which humans go about solving problems. Experienced human mathematicians prove theorems, like our group theory example, which appear to involve huge combinatorial explosions. How do they do it? In particular, I found a proof of that group theory example, which was 42 resolutions long. How did I do it?

Well I did not start from the statement of the problem as a set of 12 clauses and then resolve away. I started from my statement given in section 7.2 above, namely

Every group of exponent 2 is abelian.

This description of the problem uses a different language from that of the standard axiomatization of group theory – it suggests a different axiomatization. An axiomatization, perhaps, in which the the search tree is not so big?

Abelianness (or commutativity) is a significant concept in group theory, related, in my mind, to other concepts like *commutators* and *commutator subgroups*.

Definition 1: If G is group, a *commutator* is a term i(A)oi(B)oAoB where A and B are in G.

The *commutator subgroup*, G', of G is generated by {i(A)oi(B)oAoB for all A and B in G}.

The link between these concepts is a theorem that

G is abelian iff G' is trivial, i.e. G'={e}

This theorem in turn is related to a common technique for proving a group abelian, namely to show that G' is trivial by proving

i(a)oi(b)oaob = e for typical elements a and b in G

The 'commonness' of this technique was a piece of guidance information which caused me to select it as the technique to try first.

Not all groups are abelian, so clearly the exponent of G must be used (more guidance information). The exponent of a group is another important concept in group theory and is also related to other key concepts and theorems, in this case the key concept is that of *inverse* and the key theorem that

Every element of a group of exponent 2 is its own inverse,

i.e. X=i(X) for all X in G. If we use this theorem our goal is to prove that:

a o b o a o b = e for typical elements a and b in G

Now comes the only hard part of the proof, noticing that the exponent of G can be used again on the element a o b.

a o b o a o b = (a o b)o(a o b) = e

Now we have *a* proof, but can we translate it into a resolution proof from the clauses 1-12? To do this we must recover the elementary proof that

G is abelian if G' is trivial

and modify it.

The standard proof of this result is:

a o b = e o a o b
= b o i(b) o a o b
= b o e o i(b) o a o b
= b o a o i(a) o i(b) o a o b
= b o a o e
= b o a

Our proof above involved substituting a and b for i(a) and i(b). Doing this yields:

a o b = e o a o b
= b o b o a o b
= b o e o b o a o b
= b o a o a o b o a o b
= b o a o e
= b o a

It was fairly straightforward to transform this proof into the 10 step paramodulation proof. The trickiest bit was locating the implicit applications of associativity and inserting applications of clause 10. Then, the 42 step resolution proof was constructed by a laborious replacement of paramodulation steps by applications of equality axioms.

What can we learn from this exercise?

- –That we should look at alternative axiomatizations of mathematical theories, which utilize a language closer to the one with which we ourselves describe the problem. It may be that the search tree in this new axiomatization is smaller. But even if it is not, it will certainly be easier to express our intuitions about guiding search in such a language.

- –That guidance information is available, given a language to express it in. We saw: an example where this information was domain specific, concerning common techniques for proving that a group is abelian; and an example which was more general purpose, that one should try to use all the hypotheses of the theorem.

7.5 Alternative Axiomatization

Can we use logical notation to formalize the reasoning described in the previous section? What would that look like?

The theorem on which the whole proof was based was:

G is abelian iff G' is trivial

Clearly this is a double implication between two propositions the predicates of which are 'abelian' and 'trivial'. We will also want to represent the relationship between G and G'. Since G', the commutator subgroup, is uniquely determined by G, we should represent the relationship with a function, 'commutator-subgroup'. So the theorem can be formalized as:

abelian(G) $\longleftrightarrow$ trivial(commutator-subgroup(G)).

Our original statement also contained a definition of 'trivial'

i.e. $G' = \{e\}$.

Again this can be represented as a double implication between trivial(G') and $G' = \{e\}$. Actually there is a catch. Each group, G, has a different identity element, so e should be replaced by e(G). But this has repercussions. 'trivial' is not such a trivial concept as it at first appeared – it is dependent on a particular e and therefore on a particular group. It should have G as an additional parameter. Thus the axiomatization so far is:

$$abelian(G) \longleftrightarrow trivial(commutator\text{-}subgroup(G), G) \qquad (i)$$

$$trivial(G', G) \longleftrightarrow G' = \{e(G)\}$$

We also gave a definition of commutator subgroup above, as consisting of all commutators of a group. Can we formalize this? We will need to discuss the typical element of the commutator subgroup and how it relates to two typical elements of the group. For this we will want some functions, el and el', which take a group as parameter and represent typical elements of that group. The definition of commutator subgroup can then be represented as:

$$el(commutator\text{-}subgroup(G))$$
$$= i(el(G)) \; o \; i(el'(G)) \; o \; el(G) \; o \; el'(G)$$

Converting what we have so far into clausal form results in the clauses

1. $trivial(commutator\text{-}subgroup(G1), G1) \rightarrow abelian(G1)$.

2. $abelian(G2) \rightarrow trivial(commutator\text{-}subgroup(G2), G2)$

3. $G3' = \{e(G3)\} \rightarrow trivial(G3', G3)$.

4. $trivial(G4', G4) \rightarrow G4' = \{e(G4)\}$

5. $\rightarrow el(commutator\text{-}subgroup(G5)) =$
$$i(el(G5)) \; o \; i(el'(G5)) \; o \; el(G5) \; o \; el'(G5)$$

To this we may add our conjecture

Every group of exponent 2 is abelian

negated and put in clausal form

6. $\rightarrow$ exponent $(g, 2)$.

7. $abelian(g) \rightarrow$

Now taking 7 as top clause we can resolve it with 1 to form

8. $trivial(commutator\text{-}subgroup(g), g) \rightarrow$

and this with 3 to form

9. $commutator\text{-}subgroup(g) = \{e(g)\} \rightarrow$

At this stage we are stuck. We would like to combine 9 in some way with 5, but no resolutions or paramodulations are possible. There is a gap. If we consult the proof we are modelling we will see that the gap is filled by the knowledge that to show a group consists of one individual, show that a typical element of it is that individual, i.e.

   10. el(G10) = X10→ G10={X10}.

Now 10 resolves with 6 to give

   11. el(commutator-subgroup(g)) = e(g)

and 11 paramodulates into 5 to give

   12. i(el(g)) o i(el'(g)) o el(g) o el'(g) = e(g)

Exercise 20: Continue in this way representing the knowledge used in the proof, converting it into clausal form and trying to form a proof by resolution and paramodulation. Did you uncover any gaps? Can you fill them?

But this was not the only type of reasoning going on as we made the proof. Most of the time was spent discussing the theorems available. We said that (i) above was a theorem

   →theorem((i)).

We said (i) was a common technique for proving the abelianness of groups

   → common-technique((i), abelianness)

We gave a standard proof of (i)

   → proof((i), aob = eoaob =....= boa).

Whenever we pop up a level like this and start discussing and reasoning about the level below, it is called *meta-level* reasoning. The lower level is called the *object-level*. In chapter 12 we will discuss meta-level reasoning and how it can induce and guide a proof at the object-level.

7.6 A New Methodology

Subsequent chapters of this book will elaborate this second approximation to the automation of mathematical reasoning: we will describe some attempts to apply the methodology of the last two sections, namely to analyse the proofs of experienced human mathematicians and to try and formalize what we see there.

Thus, instead of trying to describe uniform, domain independent techniques for guiding search – techniques which are derived from work in Mathematical Logic and which have failed to produce successful theorem provers – we will seek guidance techniques in actual proofs and then try to generalize these.

In rejecting the uniform proof procedure methodology of the 1960s we must be careful not to throw out the baby with the bath water. Resolution theorem provers gained us a number of advances.

– They enabled us to specify a problem and the axioms from which its solution must be built without having to specify the solution itself.

– This was achieved with the aid of a language for the representation of knowledge, a semantics for defining the meaning of this language and a definition of proof.

– They contributed a number of procedures for manipulating this language – especially pattern matching procedures for applying the axioms to the problem and search techniques to enable the computer to change its mind and attempt a different solution.

– Occasionally we were able to prove theorems about our procedures, e.g. that they were sound or complete. When we can do it, this is a good trick. It is often a great help in correcting or improving our procedures. However, there is a danger here, of which we should beware, of designing procedures we can prove theorems about, rather than ones which are successful.

The principal deficiency of uniform proof procedures was the lack of a way of guiding the search. Control knobs were provided whereby the search could be directed. These were the selection of which literal to resolve away and the choice of which clause to resolve it with. But almost no advice was provided about how these control knobs were to be twiddled. Worst of all the control knobs were not designed to meet the needs of people who knew how they wanted the search to be directed. They were not designed at all, but emerged as a by-product of the construction of the theorem provers.

We will want to see what sort of control advice emerges from our analyses so that we can design the search control knobs ergonomically.

7.7 Summary

In this chapter we have seen that uniform proof procedures, like SL Resolution, are inadequate to prove non-trivial theorems. General purpose search guiding and pruning techniques have only a minor effect on curbing the combinatorial explosion. If we are to build expert artificial mathematicians, then we must use domain specific guidance information. A useful source of such information is the analysis of human generated proofs. The results of this methodology are the subject of part III of this book.

Part III:
Guiding Search

8
Decision Procedures for Inequalities

☐ This chapter describes decision procedures for some simple theories involving inequalities.
☐ Section 8.1 gives axioms for inequalities.
☐ Section 8.2 analyses some human proofs of inequality theorems.
☐ Section 8.3 clarifies the operations on types which occur in these proofs.
☐ Section 8.4 describes the Sup-Inf decision procedure. The various sub-sections and sub-sub-sections describe its various sub-procedures and sub-sub-procedures.
☐ Section 8.5 describes an extension to the Sup-Inf procedure, and discusses some other decision procedures.

In this chapter we adopt our new methodology; analysing the proofs of experienced mathematicians and trying to formalize what we see there. We begin with reasoning about inequalities between numbers.

8.1 Axioms for Inequalities

Some of the axioms for inequalities are closely analogous to those for equality. For instance, the analogues of reflexivity, transitivity and substitution for $\leq$ are:

Reflexivity: $\rightarrow X \leq X$

Transitivity: $X \leq Y$ & $Y \leq Z \rightarrow X \leq Z$

Substitution: $X \leq Y$ & $Z \leq W \rightarrow X + Z \leq Y + W$
$X \leq Y$ & $W \leq Z \rightarrow X - Z \leq Y - W$
etc.

There is, of course, no analogue for symmetry. Note the twist in the substitution axiom for subtraction. The versions for multiplication and division are even worse, and there are no versions at all for some functions. Of course, the existence or non-existence of substitution axioms for a function depends on its monotonicity with respect to each of its parameters, e.g. $+$ is monotonically increasing on both its parameters and $-$ is monotonically increasing on its first and decreasing on its second parameter.

There are analogues of reflexivity, transitivity and substitution for $\geqslant$, and analogues of transitivity and substitution (but not reflexivity) for $<$ and $>$. There are also hybrid versions of transitivity and substitution like:

Hybrid Transitivity: $X \leqslant Y$ & $Y < Z \to X < Z$

Hybrid Substitution: $X < Y$ & $Z \leqslant W \to X + Z < Y + W$

But these variants are not necessary if $<$, $\geqslant$ and $>$ are defined in terms of $\leqslant$, i.e.

Less Than: $X < Y \longleftrightarrow (X \leqslant Y$ & $\sim X = Y)$

Greater Than or Equal: $X \geqslant Y \longleftrightarrow Y \leqslant X$

Greater Than: $X > Y \longleftrightarrow (Y \leqslant X$ & $\sim Y = X)$

We also need an axiom to show that any two numbers are comparable, that is one is less than or equal to the other. Or in other words that the numbers are *totally ordered by* $\leqslant$.

Total Order: $\to X \leqslant Y$ v $Y \leqslant X$

Note that this axiom is a non-Horn clause, and often gives rise to arguments by cases.

The axioms given so far are true for natural, integer, rational or real numbers, but there are some additional special axioms which only hold for each kind of number. For instance, if the numbers are known to be naturals or integers then the order induced by $\leqslant$ is *discrete* and we have a more specific definition of $<$, namely

Discrete: $X < Y \longleftrightarrow X + 1 \leqslant Y$

If the numbers are naturals then 0 is a *lower bound*, i.e.

Lower Bound: $\to 0 \leqslant X$

The order induced by $\leqslant$ and $<$ on the rationals and reals is a *dense order;* between any two numbers we can always find a third. The axiom which expresses this property is called *interpolation*.

Interpolation: $X < Y \longleftrightarrow \exists Z (X < Z$ & $Z < Y)$

The Discrete axiom above is very useful for eliminating $<$ in terms of $\leqslant$. Unfortunately, it is not available for reals and rationals. To make up this difficiency we will introduce a special *infinitesimal* number, ϵ, and the axiom

Pseudo Discrete: $X < Y \longleftrightarrow X + \epsilon \leqslant Y$

An infinitesimal number is one which is smaller than any positive real number, but bigger than 0. The use of infinitesimals in analysis has recently been made respectable, by Abraham Robinson and others, but we will not be using them in a serious way, we will only use ε to keep track of which inequalities are strict and which non-strict. Similarly, we will use the numbers ±∞ in a non-strict way to keep track of terms which are unbounded.

Recalling the explosive properties of equality, exhibited in chapter 7, and the close analogies between the axioms for inequality and equality, it is clear that the above inequality axioms would also be very explosive if applied in an exhaustive way by a uniform proof procedure.

8.2 Some Human Proofs

Human proofs of theorems involving inequalities rarely involve the explicit application of these axioms. Consider the following example.

If X is a natural number then show that, $5.X < 11 \rightarrow 7.X < 16$

A 'natural' way to prove this conjecture is to assume $5.X < 11$ and to deduce $7.X < 16$ as a consequence. Of course, X is a typical natural number and we are not allowed to substitute for it. This much of the proof is modelled nicely by Resolution; we negate the conjecture and put it in clausal form to get

$\rightarrow 5.x < 11$

$7.x < 16 \rightarrow$

Note that the hypothesis is an assertion and the conclusion is a goal clause. The variable X has been translated into a Skolem constant, where it can be interpreted as a 'typical number'.

An obvious consequence of the hypothesis is that x is less than or equal to 2, since if it were 3 or greater then $5.x$ would be 15 or greater, which is not less than 11. Hence the maximum value that $7.x$ can take is 14, which is less than 16, QED.

This part of the proof seems to reason by assigning a *type* to x, i.e. establishing that x lies within some set of numbers. In this case X is assigned the type $\{0,1,2\}$, that is, X is known to be either 0 or 1 or 2. The proof then proceeds to reason about the *upper bound* of this type, i.e. to reason that X is less than or equal to 2.

Here is another example. If a, b and X are real numbers then prove that:

$a \leq 2$ & $2 \leq b \rightarrow \exists X (0 \leq X$ & $X \leq 5$ & $a \leq X)$

Once again we assert the hypothesis and deduce the conclusion, a process which can be modelled by negating the conjecture and deriving the clauses:

$\rightarrow$ a$\leq$2

$\rightarrow$ 2$\leq$b

0$\leq$X & X$\leq$5 & a$\leq$X $\rightarrow$

The goal is to find an X which lies between the maximum of 0 and a, at one extreme, and 5 at the other. By the interpolation axiom, we know this can be done provided the maximum of 0 and a is less than or equal to 5. There are two cases to consider, when 0 is the maximum and when a is the maximum. If 0 is the maximum then we are finished since 0$\leq$5. If a is the maximum then, since a cannot be greater than 2 and 2$\leq$5 then a$\leq$5 and this case is also proved. Note that the hypothesis, $\rightarrow$ 2$\leq$b is not used in the proof.

Again the proof uses types. It tries to show that X can be assigned a non-empty type, i.e. one with a lower bound less than or equal to its upper bound. To prove this it deduces a type for a and reasons about the upper bound of this type.

8.3 Types

The traditional method of representing the types of real numbers is as a pair of numbers, the upper and lower bounds, using square or round brackets according to whether the number can or cannot equal the bounds. For instance, the type of the constant a at the start of the above proof is $(-\infty,2]$, which means that $-\infty<a$ & a$\leq$2. In the case that 0 was a maximum this changed to $(-\infty,0]$, meaning $-\infty<a$ & a$\leq$0, and in the case that a was a maximum it changed to $[0,2]$

We can summarize the techniques which arose in these examples as follows:

- The assignment and/or updating of a type to constants or variables.

- The use of upper or lower bounds of these types to settle questions.

- The elimination of variables by translating questions about variables into questions about the legality of their types.

These type manipulation techniques offer an efficient alternative to resolution with inequality axioms for proving many conjectures about inequalities. Such conjectures frequently arise as subproblems when reasoning about algebra, analysis and computer programs, for instance. Type manipulation techniques have been used extensively by Woody Bledsoe and his co-workers when building theorem provers for these areas (see [Bledsoe 77]). Bledsoe's techniques have matured over the years, becoming more powerful and efficient, but also less intuitive. The rest of this chapter is devoted to a mature technique, the *Sup-Inf Method*.

8.4 The Sup-Inf Method

The Sup-Inf Method is a decision procedure for an area of real number arithmetic. A slightly modified version is also applicable to natural number and integer arithmetic. In fact, it was originally developed for natural numbers [Bledsoe 74], but does not constitute a decision procedure for the corresponding area of natural number arithmetic. The whole of real number arithmetic is known to be undecidable. A decidable subpart can be carved out by allowing only the additive functions, i.e. by excluding trigonometry, exponentiation, etc.

8.4.1 Bledsoe Real Arithmetic

A more precise definition of this decidable area, which we will call *Bledsoe Real Arithmetic*, is given below.

- *Constants:* all real numbers, e.g. 3.141....

- *Functions:* $+$ and $-$, but we will adopt multiplication by a natural number as an abbreviation for repeated addition, e.g. 3.x is an abbreviation for $x+x+x$.

- *Predicates:* $\leq$, $<$, $\geq$, $>$ and $=$.

- *Formulae:* Built from the above in the normal way, but using only universal quantifiers and only at the top level.

- *Axioms:* The equality axioms plus the inequality axioms for real numbers listed in section 8.1 above.

By substituting integers or natural numbers for reals in the first clause of the above definition we can define the similar theories of Bledsoe Integer Arithmetic and Bledsoe Natural Arithmetic. The formula

$$\forall Y \ \forall Z \ (5.Z \geq 2.Y+3 \ \& \ Z \leq X-Y \ \& \ \sim 3.X=5 \rightarrow 2.Y+1<3)$$
is a formula of all three theories.   (i)

8.4.2 An Overview of the Method

The idea of the Sup-Inf method is to show that the negation of a conjecture is unsatisfiable because one of its constants cannot be assigned a type. As in Resolution, the negated conjecture is put through a series of normal forms, the last of which assigns a type to each Skolem constant. We finish by showing that one of these types has a lower bound greater than its upper bound. The series of normal forms is outlined below.

(a) Let C be a conjecture.

(b) Form its universal closure C'.

(c)Negate C' and bring the $\sim$ sign inside the universal quantifiers, turning them into existential quantifiers, by repeatedly replacing formulae of the form $\sim\forall X$ A by $\exists X \sim A$.

(d)Replace the existentially quantified variables with Skolem constants, i.e. replace $\exists X$ A(X) by A(x), to form a variable free formula, C".

(e)Put C" in *disjunctive normal form*, i.e. transform it into a disjunction

$$D_1 \text{ v } ... \text{ v } D_n$$

where each disjunct, D_i, is a conjunction

$$L_{i1} \text{ \& } ... \text{ \& } L_{im}$$

of literals.

(f)Eliminate any literal L_{ij} which contains no Skolem constants, by arithmetic evaluation.

(g)Manipulate each D_i into a disjunction of pairs of inequalities of the form

$$k{\leq}x \text{ \& } x{\leq}l{-}\epsilon$$

say, one pair for each Skolem constant x, where k and l are real numbers or $\pm\infty$.

To prove the conjecture, C, we must show that each D_i is contradictory. We do this by showing that some L_{ij} is false at step (f) above or that there is an x such that $\sim k{<}l$, i.e. x cannot be assigned a type at step (g).

The first four steps of the above outline are the same as those for Resolution described in chapter 5, and disjunctive normal form is a dual of clausal form, described fully in chapter 15, so we will skip quickly over these steps with an example and concentrate our attention on the last one, the manipulation of the D_is to find the types of the Skolem constants.

As an example consider the conjecture (i) above. After universal closure this becomes:

$$\forall X \text{ } \forall Y \text{ } \forall Z \text{ } (5.Z{\geq}2.Y{+}3 \text{ \& } Z{\leq}X{-}Y \text{ \& } \sim 3.X{=}5 \rightarrow 2.Y{+}1{<}3)$$

Negating this and bringing the negation sign inside the universal quantifiers gives:

$$\exists X \text{ } \exists Y \text{ } \exists Z \sim(5.Z{\geq}2.Y{+}3 \text{ \& } Z{\leq}X{-}Y \text{ \& } \sim 3.X{=}5 \rightarrow 2.Y{+}1{<}3) \quad \text{(ii)}$$

Replacing the existentially quantified variables with Skolem constants gives:

$- - \sim(5.z \geqslant 2.y+ \ \& \ z \leqslant x-y \ \& \sim 3.x=5 \rightarrow 2.y+1<3)$

Putting this in disjunctive normal form yields the following sequence of steps:

$\sim[\sim(5.z \geqslant 2.y+3 \ \& \ z \leqslant x-y \ \& \sim 3.x=5) \ v \ 2.y+1<3]$

$\sim \sim(5.z \geqslant 2.y+3 \ \& \ z \leqslant x-y \ \& \sim 3.x=5) \ \& \sim 2.y+1<3$

$5.z \geqslant 2.y+3 \ \& \ z \leqslant x-y \ \& \sim 3.x=5 \ \& \sim 2.y+1<3$

This has only one disjunct, i.e. $n=1$, which is very common in practice, but in general we would have to consider a disjunction of conjunctions like the above.

Exercise 21: Apply all but the last step of the above procedure to the conjecture

$5.X<11 \rightarrow 7.X<16$

i.e. close, negate, Skolemize and put it in disjunctive normal form.

8.4.3 Assigning Types to Skolem Constants

We now consider the manipulation of the D_is, e.g.

$5.z \geqslant 2.y+3 \ \& \ z \leqslant x-y \ \& \sim 3.x=5 \ \& \sim 2.y+1<3$

to extract types for each of the Skolem constants, e.g. x, y and z. The manipulation is applied separately to each disjunct and is outlined below.

(a) Eliminate the predicates $\geqslant$, $>$, $=$ and $<$ and the connective $\sim$, so that the disjunct consists only of propositions of the form $S \leqslant T$. During this stage the disjunct may split into several disjuncts, each of which is treated separately.

(b) For each proposition, $S \leqslant T$, 'solve' it for each of the Skolem constants it contains, e.g. if $S \leqslant T$ contains the Skolem constant x then manipulate $S \leqslant T$ into the form $S' \leqslant x$ or $x \leqslant T'$, where S' and T' do not contain x.

(c) For each Skolem constant, combine together all its 'solutions' into one, written $S' \leqslant x \leqslant T'$.

(d) For each Skolem constant, manipulate its 'solution' to remove all other Skolem constants, getting a solution $k+e \leqslant x \leqslant l$, say, in which k and l are real numbers. (k,l] is the type of x.

(e) Try to find an impossible type in each disjunct.

8.4.3.1 Normalizing to Less Than or Equal

The first step is to eliminate negation and all predicates except $\leq$. This can be done in various ways, but a simple one is as follows:

(a) Eliminate $>$ and $\geq$ by replacing: all propositions of the form $S>T$ by the equivalent $T<S$; and all propositions of the form $S\geq T$ by the equivalent $T\leq S$.

(b) The only occurrences of $\sim$ signs are immediately dominating the predicates $<$, $\leq$ and $=$. They can be eliminated by replacing: all literals of the form $\sim S<T$ by the equivalent $T\leq S$; all literals of the form $\sim S\leq T$ by the equivalent $T<S$; and all literals of the form $\sim S=T$ by the equivalent $S<T$ v $T<S$. Note that in the last case a disjunction has been introduced into the conjunction. Split the conjunction into two conjunctions, by distributing v over &, and deal separately with each.

(c) Eliminate the remaining occurrences of $=$ by replacing all propositions of the form $S=T$ by the equivalent $S\leq T$ & $T\leq S$. This will merely increase the length of the conjunction.

(d) Finally, eliminate $<$ by replacing all propositions of the form $S<T$ by $S+\epsilon\leq T$, where ϵ is a new constant standing for an infinitesimal number.

The result will be a conjunction of propositions of the form $S\leq T$.

Applying these steps to our running example yields the following steps.

$5.z\geq 2.y+3$ & $z\leq x-y$ & $\sim 3.x=5$ & $\sim 2.y+1<3$

$2.y+3\leq 5.z$ & $z\leq x-y$ & $\sim 3.x=5$ & $\sim 2.y+1<3$

$2.y+3\leq 5.z$ & $z\leq x-y$ & $(3.x<5$ v $5<3.x)$ & $3\leq 2.y+1$

$(2.y+3\leq 5.z$ & $z\leq x-y$ & $3.x<5$ & $3\leq 2.y+1)$ v
$(2.y+3\leq 5.z$ & $z\leq x-y$ & $5<3.x$ & $3\leq 2.y+1)$

$(2.y+3\leq 5.z$ & $z\leq x-y$ & $3.x+\epsilon\leq 5$ & $3\leq 2.y+1)$ v   (iii)
$(2.y+3\leq 5.z$ & $z\leq x-y$ & $5+\epsilon\leq 3.x$ & $3\leq 2.y+1)$ (iv)

Exercise 22: Eliminate $\sim$ and $<$ from

$5.x<11$ & $\sim 7.x<16$

8.4.3.2 'Solving' Each Inequality

We now 'solve' each inequality for each Skolem constant it contains. 'Solving' inequalities is much like solving equations, and solving linear inequalities is particularly simple.

(a)Suppose we are solving S≤T for x.

(b)This is equivalent to 'solving' S−T≤0.

(c)Put S−T in polynomial normal form with respect to x to yield Ax+B, where A is a real number.

(d)If A is positive then the 'solution' is $x \leqslant B/A$.

(e)If A is negative then the 'solution' is $B/A \leqslant x$.

(f)If A is zero then the 'solution' is t.

The inequality 2.y+3≤5.z contains two Skolem constants: y and z. 'Solving' it for y yields the following steps:

2.y+3≤5.z

2.y+3−5.z≤0

y ≤ 5.z/2 − 3/2

'Solving' each of the inequalities in

2.y+3≤5.z & z≤x−y & 3.x+e≤5 & 3≤2.y+1

for the Skolem constants they contain yields the following 'solutions':

y ≤ 5.z/2 − 3/2 & 2.y/5 + 3/5 ≤ z &
z+y ≤ x & y ≤ x−z & z ≤ x−y &
x ≤ 5/3−e &
1≤y   (v)

Exercise 23: 'Solve' each of the inequalities in

5.x+e≤11 & 16≤7.x
for x.

8.4.3.3 Combining 'Solutions'

The 'solutions' for each Skolem constant must now be combined together into one. For instance, there are three 'solutions' for y above.

y ≤ 5.z/2 − 3/2 &
y ≤ x−z &
1 ≤ y

We can combine them into one by using the maximum lower bound and the minimum upper bound.

The only lower bound of y is 1, so this is the maximum. y has two upper bounds, namely: 5.z/2 −3/2 and x−z. What is the minimum of 5.z/2 −3/2 and x−z? We avoid the question by introducing the function *min*, on sets of numbers, and writing the combined 'solution' as:

$$1 \leqslant y \leqslant \min\{5.z/2 -3/2, x-z\}$$

Similar problems with finding the maximum lower bound are solved by using the dual function *max*.

In general the procedure to find the new lower bound for a Skolem constant x is:

(a) Collect together the lower bounds from each of the 'solutions' for x to form the set Bnds.

(b) If Bnds is the empty set, {}, then the new lower bound is $-\infty$.

(c) If Bnds is a singleton, {t}, then the new lower bound is t.

(d) If all members of Bnds are numbers then the maximum number, k, is the bound.

(e) Otherwise, the new lower bound is max(Bnds).

And a dual procedure will find the new upper bound.

Applying these procedures to the solutions (v) above yields the new upper and lower bounds given in table 8-1 below.

Table 8-1: New Upper and Lower Bounds for Skolem Constants

Skolem Constant	Lower Bound	Upper Bound
x	z+y	5/3−ϵ
y	1	$\min\{5.z/2 - 3/2, x-z\}$
z	2.y/5 + 3/5	x−y

Exercise 24: Combine together the 'solutions':

$$x \leqslant 11/5 - \epsilon \ \& \ 16/7 \leqslant x$$

to get upper and lower bounds for x. Do these bounds assign a possible type to x?

8.4.3.4 Working Out the Types

Some of these bounds are already free of other Skolem constants, e.g. the upper bound of x is 5/3−ϵ and the lower bound of y is 1. The next step is to eliminate the Skolem constants from all the remaining bounds. This step is the heart of the Sup-Inf method. The idea is as follows. Suppose the lower bound of x contains the Skolem constant y (as it does in the running example) then we replace the occurrences of y with either the lower bound or upper bound of y. This is justified by the transitive and substitution laws for ≤. If the new lower bound of x still contains Skolem constants the procedure is repeated. The main problem is that the procedure may loop, e.g. suppose the lower bound of y contains x, so that y is replaced by x and then y by x ad infinitum. We must trap this case and do something different, in fact 'solve' for x.

For instance, to eliminate z and y from the lower bound for x we generate the following steps:

$$z+y \geqslant (2.y/5 + 3/5) + 1 \tag{vi}$$
$$\text{(by substituting lower bounds of z and y)}$$
$$\geqslant (2.2/5 + 3/5) + 1$$
$$\text{(by substituting lower bound of y)}$$
$$= 12/5 \text{ (by arithmetic)} \tag{vii}$$

This assigns the type [12/5, 5/3−ε] to x, from which we can eliminate ε and rewrite as [12/5, 5/3). But this is not a possible type, since 5/3 ≤ 12/5, so x cannot have a type. This is enough to establish that (iii) is a contradiction. If we can also establish that (iv) is a contradiction then we will have shown that (ii) is a contradiction and, hence, that (i) is a theorem.

In line (vii) above we used arithmetic to evaluate a purely numeric expression. We can extend this evaluation to expressions involving ε by using the rewrite rules (see chapter 9):

$$X>0 \rightarrow X.\varepsilon => \varepsilon$$

$$0.\varepsilon => 0$$

$$X<0 \rightarrow X.\varepsilon => -\varepsilon$$

'Awkward' cases like ε−ε will never arise in the uses we make of ε.

In line (vi) above, we substituted the lower bounds for both parameters of +. This is justified by the substitution axiom for +, i.e. because + is monotonically increasing in both parameters. In the case of z−y we would substitute the lower bound for the first parameter and the upper bound for the second parameter, e.g.

$$z-y \geqslant (2.y/5 + 3/5) - \min\{5.z-3/2, x-z\}$$

since − is monotonically increasing in its first parameter, but monotonically decreasing in its second parameter. Both min and max are monotonically increasing in each member of their set parameter, so in the case of min{z, y} we would substitute the lower bounds for both members, e.g.

$$\min\{z, y\} \geqslant \min\{(2.y/5 + 3/5), 1\}$$

This procedure would loop if x had the lower bound y and y had the lower bound x−6. When trying to eliminate the y from the lower bound of x it would generate the sequence:

$$y \geqslant (x-6)$$
$$\geqslant (y-6)$$
$$\geqslant ((x-6)-6)$$
$$\ldots\ldots\ldots$$

This sequence must be trapped as soon as an x has appeared in the lower bound for x. At this stage we have the inequality:

$$x-6 \leqslant x$$

'Solving' this for x yields 'solution' t, i.e. x is unrestricted.

If x had had the lower bound y and the upper bound ∞ and y the lower bound 6−x then the procedure does terminate, but with a non-optimal lower bound, i.e.

$$y \geqslant (6-x)$$
$$\geqslant (6-\infty)$$
$$\geqslant -\infty$$

Trapping the procedure at the point

$$6-x \leqslant x$$

and 'solving' this inequality for x, yields the optimal lower bound 3.

The procedure, *Sup,* for eliminating Skolem constants from lower bounds is summarized in table 8-2. The procedure, *Inf,* for eliminating Skolem constants from upper bounds is the dual of this. Sup takes two inputs, J and H. J is the term for which a numeric upper bound is sought. It is initially a Skolem constant, but on recursive calls can become a complex term. H is a set of Skolem constants, initially empty, but taking non-empty values on recursive calls. It lists some Skolem constants which are allowed to appear in the output of Sup. The idea is that an attempt to eliminate the constants in H, using Sup, might cause a loop, so they are eliminated by another procedure, Supp, which 'solves' inequalities. Upper(J) returns the upper bound of J. Inf, Inff and Lower are the duals of Sup, Supp and Upper. Simp is a procedure for putting terms in a normal form. It distributes . over + and pulls maxs and mins out to the front, e.g. Simp applied to 4.(3+max{2,x}) is max{20, 12+4.x}.

Table 8-2: The Definition of Sup(J,H)

Condition	Action	Result
If J is a number		J
If J is a Skolem constant		
JeH		J
JéH	Let Q=Upper(J)	
	Let Z=Sup(Q,HU{J})	Supp(J,Simp(Z))
If J has form N.A where		
N is a number		
N<0		N.Inf(A,H)
N≥0		N.Sup(A,H)
If J has form N.C+B,	Let B'=Sup(B,HU{C})	
where N is a number		
and C a Skolem constant		
C occurs in B'	Let j=Simp(N.C+B')	Sup(J',H)
C does not occur in B'		Sup(N.C,H)+B'
If J has form min(S)		min{Sup(A,H): AεS}

This completes the description of the Sup-Inf decision procedure for Bledsoe Real Arithmetic.

8.5 Variable Elimination

In this section we consider how the Sup-Inf procedure can be extended to deal with a wider area of Mathematics than Bledsoe Real Arithmetic. One crucial limitation of Bledsoe Real Arithmetic was that only universal quantifiers may appear in formulae and only at the topmost level. It is possible to lift this restriction and still get a decision procedure, i.e. allowing both universal and existential quantifiers at any level still gives a decidable theory.

Table 8-3: The Definition of Supp(X,Y)

Condition	Result
If Y is a number	Y
If X≡Y	∞
Y=min(S)	min{Supp(X,A): AεS}
Y=B.X+C, where X does not occur in C	
B>1	∞
B<1	c/(1−B)
B=1	
C not a number	∞
C<0	−∞
C≥0	∞

The same is true of natural and integer arithmetic. The extension of

Bledsoe Natural Arithmetic to allow unrestricted universal and existential quantification, is called *Presburger Natural Arithmetic,* after the mathematician who first showed it decidable. Unfortunately, all known decision procedures for Presburger Natural Arithmetic are very inefficient. The best is due to David Cooper, [Cooper 72]. However, even Cooper's procedure can take on the order of $2^{2^{2^n}}$ computer steps to test a formula of size n, whereas the Sup-Inf procedure never takes more than order 2^n.

8.5.1 An Overview of the Extended Procedure

We will extend the Sup-Inf method to allow it to handle formulae of the form

$$\exists X_1 \ldots \exists X_n \, A(X_1, \ldots, X_n)$$

where A is a sentence of Bledsoe Real Arithmetic (i.e. fully closed). After negation and Skolemization these formulae differ from Bledsoe formulae only in the fact that they contain variables, so we will call the theory *Bledsoe Real Arithmetic with variables.* Note that there are no proper (i.e. non-constant) Skolem functions, because no universal quantifiers appear above any existential quantifiers. [Bledsoe & Hines 80, Shostak 79] describe extensions to handle proper Skolem functions, and hence Presburger Real Arithmetic. Shostak's extension is a decision procedure.

The key idea of the decision procedure for Bledsoe Real Arithmetic with variables, is to eliminate variables from the conjecture, using interpolation, and reduce its truth to that of a variable free formula, which is then handled by Sup-Inf. To prepare the conjecture for variable elimination it is put in clausal form; $>$, $\geq$, $\sim$, $=$ and $<$ are eliminated; and the variable to be eliminated is 'solved' for. The procedure is outlined below:

> (a) Put the conjecture C in clausal form, C', where each clause has the form:
>
> $L_1 \, \& \, \ldots \, \& \, L_n \rightarrow$
>
> where each L_i is a literal, i.e. each literal is in the antecedent.
>
> (b) Eliminate $>$, $\geq$, $\sim$, $=$ and $<$ from C', using the techniques of section 8.4.3.1, to form C".
>
> (c) Until C" contains no variables, apply the following steps.
>
> > (i) Pick a clause D from C".
> >
> > (ii) Pick a variable, X, which occurs in D.
> >
> > (iii) 'Solve' each inequality in D for X, as described in section 8.4.3.2 for Skolem constants.

(iv)Eliminate X from D, by the method described below, to form the clause D'.

(v)If D' is the empty clause then terminate with success.

(vi)Otherwise, replace D with D' in C" and repeat.

(d)C" contains no variables. Put it in disjunctive normal form and apply the later stages of the Sup-Inf method as described from section 8.4.3.1 onwards.

For instance, consider the conjecture:

$a \leqslant 2$ & $2 \leqslant b \rightarrow \exists X (0 \leqslant X$ & $X \leqslant 5$ & $a \leqslant X)$

Putting this in clausal form, with all the literals in the antecedent, yields:

$\sim a \leqslant 2 \rightarrow$
$\sim 2 \leqslant b \rightarrow$
$0 \leqslant X$ & $X \leqslant 5$ & $a \leqslant X \rightarrow$

The $\sim$ and resulting $\leqslant$ signs are easily eliminated from this to give:

$2 + e \leqslant a \rightarrow$
$b + e \leqslant 2 \rightarrow$
$0 \leqslant X$ & $X \leqslant 5$ & $a \leqslant X \rightarrow$

These clauses only contain one variable, X. This can be eliminated from the third clause, by the method described below, to give:

$2 + e \leqslant a \rightarrow$
$b + e \leqslant 2 \rightarrow$
$0 \leqslant 5$ & $a \leqslant 5 \rightarrow$

And applying the Sup-Inf method to these clauses proves the conjecture.

Exercise 25: Check this.

8.5.2 Elimination of Variables using Interpolation

It only remains to explain how a variable can be eliminated from a clause consisting of a set of 'solutions' for it, e.g. how to eliminate X from

$0 \leqslant X$ & $X \leqslant 5$ & $a \leqslant X \rightarrow$

This is done by applying the interpolation axiom

$U \leqslant W$ & $V \leqslant W \rightarrow \exists Z [U \leqslant Z$ & $Z \leqslant W$ & $V \leqslant Z]$

which gives:

$0 \leqslant 5$ & $a \leqslant 5 \rightarrow$

with unifier {0/U, a/V, 5/W, X/Z}.

In general an infinite collection of interpolation axioms is required to justify this step, but the general procedure can be defined without reference to them. The only literals in the clause D which contain X, the variable to be eliminated, will be all to the left of the → and one of:

a $\leq$ X for i=1,...,m or
X$\leq$$b_j$ for j=1,...,n

These literals are removed from D and replaced by the literals

a_i $\leq$ b_j for i=1,...,m and j=1,...,n.

8.6 Summary

The Sup-Inf method is an efficient decision procedure for Bledsoe Real Arithmetic. It can be adapted to Bledsoe Natural and Integer Arithmetic, but is not a decision procedure for these areas. The Sup-Inf method can be extended to a decision procedure for Bledsoe Real Arithmetic with variables by a technique of variable elimination. Bledsoe and Hines, and Shostak consider further extensions which allow proper Skolem functions, and hence deal with Presburger Real Arithmetic. Shostak's extension is a decision procedure for Presburger Real Arithmetic. Cooper's method is an inefficient decision procedure for Presburger Natural Arithmetic.

The Sup-Inf method works by trying to assign types to Skolem constants. It gets a 'solution' for each Skolem constant and then eliminates other Skolem constants from this solution in a recursive process which appeals to their partially assigned type boundaries. The Sup-Inf method works with the negation of the conjecture in disjunctive normal form and proves the conjecture if it is unable to assign a type to one Skolem constant in each disjunct. The method can be extended to formulae with free variables by adding a technique for eliminating such variables.

Bledsoe and Presburger Arithmetics form important subparts of many mathematical theories. They arise in Number Theory, Algebra and Program Verification. Having a decision procedure for them means that many simple lemmas can be quickly handled, freeing resources for tackling the main theorem.

Further Reading Suggestion

[Shostak 77] is a readily available description of the Sup-Inf method in a well developed state.

9
Rewrite Rules

☐ This chapter introduces rewrite rules.
☐ Section 9.1 defines the rewriting rule of inference.
☐ Section 9.2 gives examples of rewrite rules from the domains of Propositional Logic, Algebra and Peano Arithmetic.
☐ Section 9.3 discusses the termination of rewrite rule sets.
☐ Section 9.4 discusses the Church-Rosser and canonical properties of rewrite rule sets.
☐ Section 9.5 discusses strategies for applying rewrite rules.
☐ Section 9.6 discusses methods for proving a rule set Church-Rosser and canonical by showing that all its critical pairs are conflatable. This suggests a technique for making non-confluent rule sets conflatable by adding non-conflatable critical pairs as new rules (9.6.3).

When we described the standard proof of theorem (i) in section 7.4, we used the notation

$$
\begin{aligned}
a \ o \ b &= e \ o \ a \ o \ b \\
&= b \ o \ i(b) \ o \ a \ o \ b \\
&= b \ o \ e \ o \ i(b) \ o \ a \ o \ b \\
&= b \ o \ a \ o \ i(a) \ o \ i(b) \ o \ a \ o \ b \\
&= b \ o \ a \ o \ e \\
&= b \ o \ a
\end{aligned}
$$

Such chains of equalities are very common in mathematical proofs, as are chains of inequalities like the one from section 8.4.

$$
\begin{aligned}
a &\geq b + c \\
&\geq 2 + c \\
&\geq 2 + 2.b/5 + 3/5 \\
&\geq 2 + 2.2/5 + 3/5
\end{aligned}
$$

Also very common is the variant

$$
\begin{aligned}
\sim((p \ v \sim q) \ \& \sim r) &\longleftrightarrow \sim(p \ v \sim q) \ v \sim\sim r \\
&\longleftrightarrow \sim(p \ v \sim q) \ v \ r \\
&\longleftrightarrow (\sim p \ \& \sim\sim q) \ v \ r \\
&\longleftrightarrow (\sim p \ \& \ q) \ v \ r
\end{aligned}
$$

or with '⟷' replaced by 'iff'.

Why is this notation so common? Perhaps because the inference system being used is to *rewrite* the expression on the left of the $=$ or $\longleftrightarrow$ sign into the expression on the right. The notation neatly captures the resulting, gradual transformation of the initial expression into the final one. It is called the application of *rewrite rules*. Not only are rewrite rules a common inference system in everyday mathematics but they can also be a very efficient computational technique, that is they can involve very little search. We will devote this chapter to a more detailed study of them, including a discussion of some interesting theoretical results.

9.1 What are Rewrite Rules?

Rewrite rules then are a set of ordered pairs of expressions. We will write a typical such pair (or rule) as

lhs $=>$ rhs.

There will usually be some kind of *similarity relation* between lhs and rhs. This might be equality, inequality, double implication or sometimes only implication (i.e. lhs $\rightarrow$ rhs).

> *Definition 1:* If, by ignoring the order of the pairs and allowing rewriting with rhs $=>$ lhs as well lhs $=>$ rhs, one expression can be rewritten into another, then the two expressions are said to be *similar* with respect to the set of rules.

In general, two expressions can be similar, without it being possible to rewrite one into the other, because a similarity relation may be available as rule lhs $=>$ rhs but not as rule rhs $=>$ lhs.

To apply these rules to an expression we need the *rewriting rule of inference*. Let exp[sub] be the expression we are trying to rewrite where sub is a distinguished subexpression of exp. The rewriting rule is:

exp[sub]
lhs $=>$ rhs
_____

exp[rhsφ]
where φ is a most general substitution such that lhsφ$\equiv$sub

That is, lhs is matched against some subexpression of exp and this subexpression is replaced with rhs. exp[rhsφ] is called a *rewriting* of exp[sub].

If the similarity relation between lhs and rhs is $=$ then the rewriting rule of inference is a simplification of paramodulation. One of the main ways in which rewriting is a simplification of paramodulation is that the substitution φ is only applied to lhs and not to sub. We call this restriction of unification, *one way matching.* This is not a serious limitation in practice since in most applications the expression to be rewritten is variable free. Hence variables will never appear in the expression unless they are introduced by rules like

$$0 => X.0$$

which contain variables in the right hand side which are not in the left hand side. With full unification such rules would cause infinite chains of rewritings, e.g.

$$0 = X.0$$
$$= (X.0).0$$
$$= ((X.0).0).0$$
$$......$$

For this reason we exclude them and allow only rewrite rules in which each variable on the right hand side appears on the left hand side. This is not a serious limitation in practice, since we seldom want to use such rules.

9.2 Some Sample Rewrite Rule Sets

9.2.1 Literal Normal Form

Many examples of the use of rewrite rules can be found in chapter 15. For instance, in section 15.4 we use the following rewrite rule set for literal normal form:

$$\sim\sim A => A$$
$$\sim(A \text{ v } B) => \sim A \text{ \& } \sim B$$
$$\sim(A \text{ \& } B) => \sim A \text{ v } \sim B$$

Suppose these rules are to be applied to the formula,

$$\sim(p \text{ v } \sim q) \text{ v } r$$

There is only one subexpression of this formula which unifies with the left hand side of any rule. The subexpression is

$$\sim(p \text{ v } \sim q),$$

the rule is

$$\sim(A \text{ v } B) => \sim A \text{ \& } \sim B$$

and the matching substitution

$$\{p/A, \sim q/B\}$$

Hence the rewriting obtained by applying the rule is

$(\sim p \& \sim\sim q) \vee r$ (i)

to which a further rule will apply.

Exercise 26: Apply another rule to (i). Say what exp, sub, lhs, rhs and φ are.

9.2.2 Algebraic Simplification

A major application of rewrite rules is in the simplification of algebraic expressions. The problem of simplifying algebraic expressions arises as a frequent subproblem in the provision of the 'mathematicians aid' computer programs mentioned in the introduction and to be explored in more detail in chapters 12 and 18. When such a program integrates or differentiates an expression the result may look something like

$a^{2.0}.5 + b.0$ (ii)

which could stand considerable simplification.

This simplification is effected by a set of rewrite rules, like

1. $X.0 \Rightarrow 0$
2. $1.X \Rightarrow X$
3. $X^0 \Rightarrow 1$
4. $X+0 \Rightarrow X$

Applying these to expression (ii) we can rewrite it as follows.

$$
\begin{aligned}
a^{2.0}.5 + b.0 &= a^0.5 + b.0 && \text{(by 1.)} \\
&= 1.5 + b.0 && \text{(by 3.)} \\
&= 5 + b.0 && \text{(by 2.)} \\
&= 5+0 && \text{(by 1.)} \\
&= 5 && \text{(by 1.)}
\end{aligned}
$$

This is not the only way to rewrite the expression. We could have started by rewriting b.0 to 0, say. In fact, as with the other rules of inference we have met, resolution and paramodulation, we may have several choices at every stage. Thus the rewriting rule also defines a search tree. The beginning of this for the above example is shown in figure 9-1.

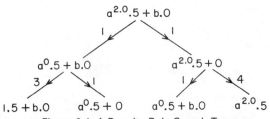

Figure 9-1: A Rewrite Rule Search Tree

We have seen how expressions can be put in normal form by the repeated application of rewrite rules. The final normal form could be any one of the rewritings labelling the tips of the complete search tree.

9.2.3 Evaluation

As a final example, numerical calculations can be done using rewrite rules. We make rewrite rules from the recursive definitions of the arithmetic functions.

1. $X+0 => X$
2. $X+s(Y) => s(X+Y)$
3. $X.0 => 0$
4. $X.s(Y) => X.Y + X$

and apply these to the arithmetic expression to be evaluated.

$$
\begin{aligned}
s(0).s(s(0)) &= s(0).s(0) + s(0) & \text{(by 4.)} & \quad \text{(iii)}\\
&= [s(0).0 + s(0)] + s(0) & \text{(by 4.)} \\
&= [0 + s(0)] + s(0) & \text{(by 3.)} \\
&= s(0+0) + s(0) & \text{(by 2.)} \\
&= s(0) + s(0) & \text{(by 1.)} \\
&= s(s(0) + 0) & \text{(by 2.)} \\
&= s(s(0)) & \text{(by 1.)}
\end{aligned}
$$

If the expressions being manipulated are *non-symbolic*, i.e. they are terms made from the constant 0 and the functions s, + and ., then it is easy to prove by induction that the application of these rewrite rules will eventually terminate with a *specific number*, i.e. a term made only from 0 and s. We call this process the *evaluation* of the non-symbolic expression.

If the expression being manipulated is *symbolic*, i.e. if it contains a Skolem constant or function, then the application of these rewrite rules will still terminate, but not necessarily with a specific number. For instance, applied to s(a).s(b) the process produces successively,

$$
\begin{aligned}
s(a).s(b) &= s(a).b + s(a) & \text{(by 4.)} \\
&= s(s(a).b + a) & \text{(by 2.)}
\end{aligned}
$$

and then terminates. When evaluation is applied to symbolic expressions it is usually called *symbolic evaluation*.

Evaluation and symbolic evaluation can be used, not just for numeric calculations, but whenever we have a theory in which functions are defined by recursive equations. In chapter 11 we will see how it can be applied to a theory of lists.

9.3 Termination

A nice property of many sets of rewrite rules, including those above, is that the application of rules to expressions cannot go on forever – it will eventually *terminate.*

The termination of the rules for literal normal form (our first example above) is proved in section 15.4. We define a numerical function on a formula, called its *load,* which measures the total size of all formulae dominated by ~ signs and show that it decreases each time a rule is applied. Since the load can never be negative there must come a time when no further rewrite rules can be applied.

Similar arguments can be made for the other two sets above. In the case of the simplification rules each application of a rule will produce an expression with a smaller expression tree (i.e. fewer nodes) than the one before.

Exercise 27: Show that the application of the rule:

$$X.Y + X.Z => X.(Y+Z)$$

will terminate.

Why is termination an important property of a set of rewrite rules?

If a set of rules always terminates and only finitely many rules can be applied to any expression, then any search tree generated must be finite. This means that if we want to know whether one expression will rewrite into another we can search the whole tree without fear that the process will go on forever (although, it may still go on for a very long time).

To prove termination of a set of rules we proceed as above. We find some non-negative numeric measure of the expression, that strictly decreases every time a rule is applied, and then we reason that the rules must stop applying at some stage.

A set of rules will fail to terminate when they contain rules like commutativity

$$XoY => YoX$$

which cause the process to loop; or when an equation is used both ways round, e.g.

$$X.(Y+Z) => X.Y + X.Z$$
$$X.Y + X.Z => X.(Y+Z)$$

which has the same effect; or when the left hand side of a rule is a subexpression of the right hand side, e.g.

$$X => X.1.$$

9.4 Other Important Properties

Termination is one important property of a set of rewrite rules. Other important properties are being *Church-Rosser* and being *canonical*.

> *Definition 2:* A set of rules is Church-Rosser when similar expressions have a common rewriting.

Where 'similar' is used in the technical sense defined in section 9.1. The property is named after the two mathematicians who first investigated it.

Suppose the rewrite rules were produced from a set of equations. If they are Church-Rosser then we have not lost anything by making them into rewrite rules and applying them only left to right. Any two expressions which could have been shown equal with the aid of the equations can now be shown equal by developing their respective rewriting search trees, until a common rewriting is uncovered.

If the set of rules is also terminating and finite then these search trees are finite and can be completely developed. Hence, this technique constitutes a decision procedure for deciding whether two terms are equal.

> *Definition 3:* A set of rules is canonical when all the normal forms of each expression are identical.

That is, it does not matter how you go about applying the rules to an expression, you will get the same result. The common normal form is called the *canonical form* of the starting expression at the root.

When a set of rules is canonical it is not necessary to search at all. Any choices can be made, in the confidence that the result will be the same in the end.

We will address the question of how rewrite rule sets may be proved to be canonical and Church-Rosser in section 9.6 below.

9.5 Applying Rewrite Rules

Even when a set of rules is canonical we may still want to exercise our choices carefully, in order to minimise the amount of computational effort required to find the canonical form. We may want to explore only the shortest branch. If the rules are non-terminating we may want to avoid infinite branches.

To help us do this various rewrite rule application procedures are available. A careful inspection of the rules to be applied may reveal that one of these is appropriate to use.

9.5.1 Inside Out Application

One of the simplest and most popular application procedures is to proceed from the inside and work outwards as we did when calculating the value of s(0).s(s(0)) in example (iii) above. For instance, the expression

$$[s(0).0 + s(0)] + s(0)$$

can be rewritten in three ways: rule 3 can be applied to s(0).0; rule 2 can be applied to [s(0).0 + s(0)] or rule 2 can be applied to the whole expression. In (iii) we chose the first possibility, that is we chose to rewrite the leftmost/innermost subexpression which *could* be rewritten. The same criterion determined the other choices in (iii).

This procedure is usually called, *call by value*, by analogy with the procedure of the same name used by many programming languages to evaluate computer programs. Under this analogy the 'call' is the rewriting process and the 'value' is the canonical form. Call by value can be summarized as follows.

> To rewrite an expression using call by value
> 1. Search the expression tree of the expression
>   by left-first/depth-first search.
> 2. At each node:
>   If the node is a tip then back up.
>   If subtrees below the node are completely rewritten
>   (i.e their values are known)
>   and a rewrite rule will apply to the subexpression
>   dominated by this node
>   then apply this rule and exit the procedure
>   Otherwise continue searching.

9.5.2 Outside In Application

In some circumstances call by value is hopelessly longwinded. Consider what happens when it is used to apply the literal normal form rewrite rules to $\sim\{\sim(p \vee \sim q) \& r\}$.

$\sim(p \vee \sim q)$ is first rewritten to $\sim p \& \sim\sim q$

then $\sim\sim q$ is rewritten to q

only then can the outermost $\sim$ be brought inside, producing $\sim(\sim p \vee q) \vee \sim r$

The earlier work on $\sim(p \vee \sim q)$ must then be undone with $\sim(\sim p \vee q)$ being rewritten as $\sim\sim p \& \sim q$ and then $p \vee \sim q$.

For the literal form rules an application procedure that starts from the

outside and works inwards, sweeping the ~s before it, is more appropriate. This would put ~{~(p v ~q) & r} in literal form in two
rewritings.

$$\sim\{\sim(p\ v\ \sim q)\ \&\ r\} \longleftrightarrow \sim\sim(p\ v\ \sim q)\ v\ \sim r$$
$$\longleftrightarrow (p\ v\ \sim q)\ v\ \sim r$$

This application is called *call by name*, again by analogy with the evaluation
procedures of programming languages. The 'name' here is the subexpression being rewritten. Call by name can be summarized as

    To rewrite an expression using call by name
    1. Search the expression tree of the expression by
           left-first/depth-first search.
    2. For each node encountered
        If the node is a tip then back up.
        If a rewrite rule will apply to the
           subexpression dominated by the node
        then apply it and exit the procedure.
        Else continue the search.

    Call by name is superior to call by value in that it will sometimes avoid an
infinite branch that call by value will plunge down. Consider the rules:

    a => a.1
    X.0 => 0

applied to the expression a.0. Call by value will generate the non-
terminating sequence

    a.0 = a.1.0
       = a.1.1.0
      ....

whereas call by name will immediately rewrite a.0 to 0 by applying the
second rule.

    The various rewrite rule systems considered in chapter 15 can all be
sensibly applied using either call by value or call by name, with call by name
being superior in each case. In chapter 12 we consider more subtle, selective
applications of rewrite rules, designed to cut down further the amount of
search.

9.6 Proving Rules Canonical and Church-Rosser

We will look at only one technique for proving a set of rules canonical or
Church-Rosser: that is to prove a property which implies both of them: the

property of *confluence*.

Definition 4: A set of rules is confluent if whenever an expression, exp, can be rewritten in two different ways, say to int1 and to int2, then int1 and int2 can both be rewritten to some common rewriting, comm.

We will say that exp is *unambiguous* and that int1 and int2 are *conflatable*. This situation is summarized in figure 9-2, where the arrows labelled by *s indicate any number of rewritings (including 0).

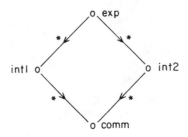

Figure 9-2: Definition of Confluence

The algebraic simplification rewrite rules of section 9.2.2 are confluent. For instance, $a^{2.0}.5 + b.0$ can be rewritten to $1.5 + b.0$ or to $a^{2.0}.5$, but both of these can be subsequently rewritten to the common rewriting, 5.

It is easy to see that a confluent set of rules is canonical.

Theorem 5: A confluent set of rules is canonical

Proof: We assume the set of rules is confluent and try to show it canonical. Consider two of the normal forms of some expression, exp. Since these normal forms are rewritings of exp then they can play the roles of int1 and int2 above. Hence they conflate to some common rewriting, comm. But these are normal forms, i.e. they cannot be further rewritten! So they must already be identical. So all normal forms are identical and the rules are canonical.   QED

Showing that a confluent set is also Church Rosser is only slightly more complicated.

Theorem 6: A confluent set of rewrite rules is Church-Rosser

Proof: Suppose exp'and exp" are two expressions which can be shown similar by using the rules in a confluent set either way round. The proof of this will be a chain of similar expressions consisting of runs of applications left to right and runs of applications right to left.

The chain showing exp' similar to exp" is diagrammed in figure 9-3.

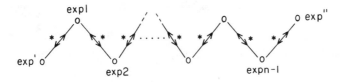

Figure 9-3: The Chain Showing exp' and exp" are Equivalent

We must show that exp' and exp" have a common rewriting.

We will construct this common rewriting by developing the chain downwards. The construction is shown in figure 9-3 below. It uses induction on the number of runs in the chain. We assume that the common rewriting can be constructed when the chain has n−1 runs and prove it in the case of n runs.

The case of 0 runs is trivial, since exp' and exp" are then identical and thus conflatable. The case of 1 run is also trivial, since either exp' or exp" is rewritable into the other and, hence, shares all its normal forms.

For the case of n ≥ 1 consider the triangle formed by exp', exp1 and exp2. Let these play the roles of int1, exp and int2, respectively in the confluence figure. From this we can see that exp' and exp2 conflate into a common expression, comm1.

Now comm1 can be shown similar to exp" by using rewrite rules either way round and in a chain of n runs, since the steps from comm1 to exp2 and exp2 to exp3 merge into one run of right to left applications. Hence by the induction hypothesis, comm1 and exp" conflate into a common rewriting comm. But comm is also a common rewriting of exp', so exp' and exp" conflate.

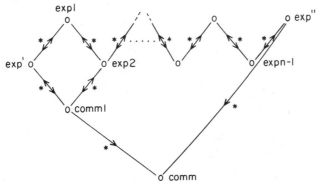

Figure 9-4: The Construction of a Common Rewriting

If the first run from exp' is left to right instead of right to left then exp1 can immediately be used in the role of comm1. comm may not be a common normal form itself, because it may be possible to rewrite it further, but any normal form of comm will be a common normal form of exp' and exp".

QED

Thus if we can show that a set of rules is confluent, we know it is also canonical and Church-Rosser. But how can we show a set of rules is confluent?

9.6.1 Local Confluence

We would like to be able to test for confluence by applying some simple procedure to the rules themselves. To show how this can be done we will gradually restrict the notion of confluence, by making it more and more local.

We start by defining the notion of *local confluence*. A set of rules is locally confluent if whenever an expression, exp, has two *immediate* rewritings, int1 and int2, then int1 and int2 conflate to a common rewriting, comm. This situation is diagrammed in figure 9-5

The algebraic simplification rules of section 9.2.2 are locally confluent. For instance, $a^{2.0}.5 + b.0$ can be immediately rewritten to both $a^0.5 + b.0$ and $a^{2.0} + 0$, but these can be subsequently rewritten to the common rewriting, 5.

Obviously a confluent set of rules is locally confluent, but is a locally confluent set always confluent?

The answer is yes, if the set is also terminating.

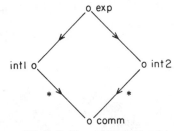

Figure 9-5: Definition of Local Confluence

Theorem 7: A terminating, locally confluent set of rewrite rules is confluent

Proof: We assume that the set of rules is locally confluent and terminating and try to show it is confluent.

Since the rules are terminating, all branches are finite so we can use induction on the structure of the search tree. We will show that if any rewritings of an expression, exp, are unambiguous then exp is unambiguous, from which we may deduce that all expressions are unambiguous and the set of rules is confluent. This proof technique is called noetherian induction in [Huet 77].

Assume that the expressions int1 and int2 are rewritings of exp. We need to show that int1 and int2 are conflatable. If either int1 or int2 is identical to exp then they are trivially conflatable, so we assume that at least one rewrite rule application is involved in each case.

Let exp1 be the first rewriting on the path to int1 and exp2 be the first rewriting on the path to int2. The situation is summarized in figure 9-6. Since the rule set is locally confluent, exp1 and exp2 must conflate to some common rewriting, comm'. But exp1 is a rewriting of exp, so by the induction hypothesis int1 and comm' must conflate to some common expression, comm1. Similarly, by the induction hypothesis int2 and comm' must conflate to some common rewriting comm2. But comm' is also a rewriting of exp, so invoking the induction hypothesis once again, comm1 and comm2 must conflate to some common rewriting, comm. Finally, comm is a common rewriting of int1 and int2, so they conflate, hence exp is unambiguous and by induction the rule set is confluent.   QED

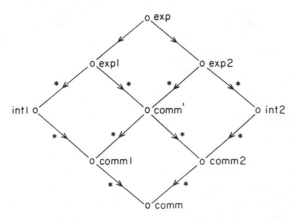

Figure 9-6: Local Confluence Implies Confluence

9.6.2 Critical Pairs

When is a set of rules locally confluent?

Let us start by looking to see how choices arise during rewriting. Clearly this happens when more than one rule applies to the current expression or one rule applies in more than one way. There are two cases to consider:

1. Either the rule applications are to totally different parts of the expressions, e.g. $X.0 \Rightarrow 0$ applies to the first part of $2.0 + 3.1$ and $Y.1 \Rightarrow Y$ applies to the second.

2. Or one rewritten subterm is totally enclosed in the other rewritten subterm, e.g. $X.0 \Rightarrow 0$ applies to the whole of $(3.1).0$ whereas $Y.1 \Rightarrow Y$ applies only to the part in parenthesis,

Of course, the two parts may be equal, as when the rules $X^0 \Rightarrow 1$ and $0^Y \Rightarrow 0$ are applied to 0^0. What cannot happen is that the two parts should intersect without total enclosure by one or the other. This follows from the fact that expressions are trees.

In the first case local confluence is retained, because the application of the rules is independent. They can be applied in any order and the result will be the same.

In the second case the application of one rule may prevent the application of the other. We must then examine the results of the two rule applications, the int1 and int2 of figure 9-5, to see whether they are conflatable. Fortunately, it is not necessary to consider all the possible expressions that the two rules may apply to, i.e. all possible 'exp's. It is possible to tell whether two rules will apply to overlapping parts of the same expression, and what the resulting expressions will look like, by examining the rules themselves. From the two rules which produce int1 and int2 we can form two expressions called a *critical pair*. The pair <int1,int2> is an instance of the critical pair and int1 and int2 will be conflatable if the critical pair is conflatable.

Suppose two rules,

lhs1 $\Rightarrow$ rhs1 and lhs2 $\Rightarrow$ rhs2,

apply to the subexpressions, sub1 and sub2, respectively of the expression, exp, then sub1 is an instance of lhs1 and sub2 is an instance of lhs2. The case we are interested in is when one of the subexpressions is totally contained in the other, say sub2 is totally contained in sub1. In this case some subexpression, bit1, of lhs1 is unifiable with lhs2 with most general unifier, say φ, so we can cover all cases of multiple overlapping application by seeing whether the left hand sides of any rules will unify with subexpressions of the left hand sides of other rules and seeing whether the resulting critical pairs conflate. If lhs1[bit1] indicates the left hand side of the rule, with the distinguished subexpression bit1, then the critical pair is <rhs1φ,lhs1[rhs2]φ>.

For instance, the possibility of overlapping applications of $X.0 \Rightarrow 0$ and $Y.1 \Rightarrow Y$ is betrayed by the fact that X unifies with Y.1 with most general unifier {Y.1/X}. The critical pair will then be <0, Y.0>. In our example the role of the overlapping instances, sub1 and sub2, was played by (3.1).0 and 3.1 respectively, and the role of the different rewritings, int1 and int2, was played by 0 and 3.0.

Exercise 28: Find a subexpression of Y.1 which will unify with X.0. Use this unification to construct another term which the rules above will both apply to.

Similarly, the fact that $X^0 \implies 1$ and $0^Y \implies 0$ both apply to 0^0 is shown by the fact that X^0 and 0^Y unify with most general unifier $\{0/X, 0/Y\}$. Here the critical pair is $<1,0>$

Note that the same rule may apply to two different overlapping subexpressions e.g. $X.0 \implies 0$ will apply to (3.0).0 to produce either 0 or 0.0. Thus when seeking all possible multiple overlapping applications we must also try unifying the left hand side of a rule with subexpressions of itself (after first standardizing apart the variables in the two copies). In this case X1.0 unifies with X2 with most general unifier $\{X1.0/X2\}$ and the critical pair is $<0, 0.0>$.

These observations can be summarized in a theorem.

Theorem 8: If all the critical pairs of a set of rules are conflatable then the set of rules is locally confluent.

Proof: Consider a situation when two rules

1. lhs1 $\implies$ rhs1 and 2. lhs2 $\implies$ rhs2

apply to the subexpressions sub1 and sub2 of exp, producing the rewritings exp1 and exp2, respectively. As noted above, if sub1 and sub2 do not overlap then rule 1 can be applied to exp2 and rule 2 to exp1 to yield the same result. Thus exp1 and exp2 conflate in this case and we only need consider the cases when sub2 is contained in sub1 and vice versa. These are symmetric, so we consider only the first case, i.e. when exp has the structure, exp[sub1[sub2]].

In this case exp1 and exp2 will be exp with sub1 and sub2 replaced by some instance of rhs1 and rhs2, i.e. some instances of exp[rhs1] and exp[sub1[rhs2]]. In fact, since sub1 and sub2 are instances of lhs1φ and lhs2φ the new expressions will actually be instances of exp[rhs1φ] and exp[sub1[rhs2φ]], see figure 9-7.

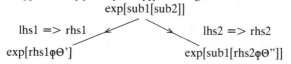

exp[sub1[sub2]]

lhs1 $\implies$ rhs1   lhs2 $\implies$ rhs2

exp[rhs1φΘ']   exp[sub1[rhs2φΘ"]]

Figure 9-7: Overlapping Rewritings of an Expression

Since bit1φ and lhs2φ are identical and sub2 is an instance of rhs2 the second of these expressions is also an instance of exp[lhs1[rhs2]φ]. Thus we need only show that the two expressions rhs1φ and lhs1[rhs2]φ are conflatable. These expressions form the critical pair <rhs1φ,lhs1[rhs2]φ>. Since every critical pair is conflatable the theorem is proved.   QED

The usual way to show that a critical pair is conflatable is to show the equivalent result that they have a common normal form. Each element of the pair should only have one normal form, otherwise we have a counterexample to the rule set being canonical.

Thus we have a way to show that a set of rules is canonical and Church-Rosser.

1. Show that the set of rules is terminating, using the method outlined in section 9.3, and that each expression can only be rewritten in a finite number of ways. If the rules are non-terminating or infinite branching then fail.

2. Find all critical pairs of the rules.

3. Derive all normal forms of each element of each pair. (Since the rules are terminating this will not take forever.)

4. If any element has more than one normal form then fail. (This test can be interleaved with the previous step for greater efficiency.)

5. If any pair have different canonical forms then fail.

6. Otherwise the critical pairs all conflate. Hence the rules are locally confluent. Since they are also terminating they are confluent. Therefore they are both canonical and Church-Rosser.

9.6.3 Improving Non-confluent Rule Sets

Even if this test should fail, all is not lost. If the test fails at step 4 or step 5 then we have two similar expressions, either two normal forms of an element or the canonical forms of a critical pair, which are not conflatable. Suppose the two expressions are nf1 and nf2, we can make them conflatable by adding a new rule,

either nf1 => nf2 or nf2 => nf1

to the rule set. nf1 and nf2 have been shown similar, since they are both normal forms of the same expression, so both the above rules are correct provided the original rule set was correct.

Naturally, this is not the end of the matter; the resulting set is not necessarily confluent.

1. There may be other non-conflatable pairs of expressions which must also be added as new rules.

2. The new rules will give rise to new critical pairs which must now be tested for conflatability.

3. Worst of all the new rule will have upset the termination proof in step 1, which must now be redone. It is possible that it cannot be redone – the set may now be non-terminating! If this happens this process of improving the rule set must come to an end.

However, if all goes according to plan – the set stays terminating and we run out of non-conflatable pairs – then the resulting rule set will be confluent.

This method of improving rewrite rule sets and possibly turning them into confluent sets is due to Knuth and Bendix [Knuth and Bendix 70]. They applied it, in various ways, to the theory of groups. For instance, starting with a non-confluent set of rules based on three group axioms: the left identity, left inverse and associative axioms; their computer program added 17 new rules until a confluent set was obtained. Several of these new rules were of interest as theorems of group theory, e.g. the right identity and right inverse axioms were generated together with

$$i(i(X)) = X \text{ and } i(XoY) = i(Y)oi(X).$$

Ten of the twenty rules were sufficient to solve the word problem for a free group with no relations. This *word problem* is to find a decision procedure to decide whether any two variable free terms are equal, according to the axioms of group theory alone. A variable free term for a free group is a term composed of only o, i and e. The decision procedure is to see if the two variable free terms have the same canonical form under this set of ten rules.

9.7 Summary

Rewrite rules can be used to put expressions in normal form and to prove expressions equal, equivalent or similar in some other way. Their application represents a powerful method of mathematical reasoning, because they can sometimes overcome the combinatorial explosions caused by other uniform proof procedures, e.g. resolution. The desirable properties of rewrite rule sets which reduce the combinatorial explosion are: termination, being Church-Rosser and being canonical. Sets of rules can be shown to terminate with the aid of a numerical function of the expression, which decreases each time the expression is rewritten. Sets of rules can be shown to be Church-Rosser and canonical by looking at critical pairs formed from pairs of rules. Sets of rules can be applied using: call by value, call by name or by selective application.

Further Reading Suggestions

[Huet 77] is a mathematically intense account of several theorems on confluence, including those given above. [Huet & Oppen 80] is a survey of work on equational systems, but it is also quite a demanding account.

10
Using Semantic Information to Guide Proofs

☐ This chapter describes the use of models to guide Horn clause proofs, illustrated mainly on the domain of Euclidean Geometry.

☐ Section 10.1 discusses the formalization of Euclidean Geometry as a set of Horn clauses.

☐ Section 10.2 describes semantic checking, the use of a diagram to prune the search tree.

☐ Section 10.3 describes the use of a diagram to suggest constructions in geometric proof.

☐ Section 10.4 shows that a diagram is a model of the axioms and hypotheses of a geometry conjecture.

☐ Section 10.5 generalizes semantic checking and applies it to a conjecture in Arithmetic.

☐ Section 10.6 explains why semantic checking does not generalize to non-Horn clauses.

☐ Section 10.7 proves that semantic checking preserves the completeness of Horn clause proofs.

When humans try to prove theorems they are not guided solely by the syntactic properties of the formulae they are manipulating. They have some idea of what it all *means* (whatever that means?) and they can bring this knowledge to bear.

Even in the original formal mathematical theory, Euclidean Geometry, we have a diagram to guide us. The diagram even has a place in the official layout of the theorem statement.

Note that the conventions for upper and lower case are rather different, in the official layout of Euclidean geometry theorem statements, than the one we have been using. We will have to alter the official layout to correspond to our convention.

Statement The angle bisector is equidistant from the rays of the angle.

Diagram

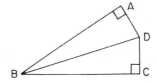

Given
1. Seg DB bisects Angle ABC
2. Seg DA⊥Seg BA
3. Seg DC⊥Seg BC

Required to Prove
4. Seg AD = Seg CD

The observation that humans usually use diagrams when trying to prove Euclidean Geometry theorems, led Herbert Gelernter, to try to build a theorem proving program which could use a diagram to guide the proof [Gelernter 63, Gelernter et al 63]. Gelernter called his program the *Geometry Machine.*

In this section we will make a rational reconstruction of the Geometry Machine. That is, we will explain its main ideas using the terminology developed in the earlier part of this book rather than the terminology used by Gelernter. We will sometimes gloss over or ignore aspects of the Geometry Machine which are incidental to our main theme of how the diagram was used to guide the search. Defying historical niceties we will sometimes refer to this rational reconstruction as the Geometry Machine.

10.1 Formalizing Geometry

Before we can discuss how a diagram might guide the search for a proof we must have a search to guide. This means choosing some inference system, and designing some formalism in which axioms and conjectures may be expressed.

The following, functions, predicates, axioms, etc are adapted from [Gelernter 63] and from [Gilmore 70]. Some were suggested by the standard abbreviations of Euclidean Geometry, some, especially the more basic ones, were invented from scratch.

One of Gelernter's goals, when designing the Geometry Machine was for it to produce proofs comparable to the normal human ones. This led him to reject the standard axiomatizations of Geometry, e.g.the ones due to Tarski, Hilbert and Forder. Instead he tried to capture, in axioms, the laws assumed by the school student. This led to, a somewhat ad hoc, highly redundant, rather large, but psychologically plausible list of axioms. It would be tedious to list them all here. Instead we give a small sample, including those needed to illustrate the use of the diagram to guide the proof.

Paul Gilmore, in an analysis of the Geometry Machine [Gilmore 70] divides the axioms into three classes.

(a) Basic axioms relating points to the higher level concepts of: line segments, angles, triangles, etc. These include such definitions as:

If A and B are distinct points, and B and C are distinct points, and A and C are distinct points, and A, B and C are not collinear, then ABC is a triangle.

We will formalize this as:

$$\text{distinct}(A,B) \ \& \ \text{distinct}(B,C) \ \& \\ \text{distinct}(A,C) \ \& \ \text{notcoll}(A,B,C) \\ \rightarrow \text{is-triangle}(A,B,C).$$

(b) Axioms expressing symmetries of these predicates and functions, e.g.,

If segment AB equals segment DC then segment AB equals segment CD.
We will formalize this as:

$$\text{seg}(A,B) = \text{seg}(D,C) \rightarrow \text{seg}(A,B) = \text{seg}(C,D). \tag{i}$$

(c) The more familiar axioms expressing equality between angles and line segments, congruence between triangles, parallelism between line segments, etc. e.g.

Corresponding sides of congruent triangles are equal,
formalized as:
$$\text{tri}(A,B,C) \equiv \text{tri}(D,E,F) \rightarrow \text{seg}(A,B) = \text{seg}(D,E), \tag{ii}$$

and

Two triangles are congruent if they have two equal angles and a corresponding equal side,

formalized as:

$$\text{angle}(A,C,B) = \text{angle}(D,F,E) \ \& \tag{iii} \\ \text{angle}(C,A,B) = \text{angle}(F,D,E) \ \& \\ \text{seg}(B,C) = \text{seg}(E,F) \rightarrow \text{tri}(A,B,C) \equiv \text{tri}(D,E,F).$$

Gilmore also points out that all of these axioms are Horn clauses. This observation has important consequences for the use of the diagram, as we will see below.

10.2 Geometric Proofs

Lush Resolution (see section 6.3) produces proofs very like those of the

Geometry Machine. So using Lush Resolution, the above axioms and a formalization of a conjecture, it is possible to produce some automatic geometry proofs – for instance, a proof of the theorem described above that

The angle bisector is equidistant from the rays of the angle.

The formalization of this conjecture, negated and put in clausal form is:

1. → seg(d,b) bisects angle(a,b,c).
2. → seg(d,a) ⊥ ~seg(b,a).
3. → seg(d,c) ⊥ ~seg(b,c).
4. seg(a,d) = seg(c,d) →.

Figure 10-1 contains a fragment of the search tree generated by taking the goal clause, 4, as top clause and applying the axioms from (b) and (c) above to it using Lush Resolution.

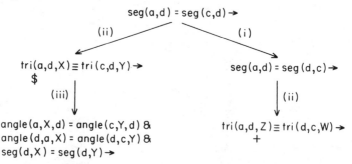

Figure 10-1: A Fragment of the Search Tree for a Simple Geometry Theorem

Consider the two 'congruence' goals, marked $ and + in figure 10-1. (NB Just the goals, i.e. the antecedent of the clause, not the whole clause). What could be substituted for the variables, X, Y, Z and W so that these goals were provable? If we restrict the possible substitutions to points occurring in the diagram the possibilities are extremely limited.

 – Substituting b for X and for Y will make $ true in the diagram and hence quite likely provable by the program.

 – No substitutions of points for Z and W will make + true in the diagram; to find suitable points would require making a construction.

The Geometry Machine used such reasoning to guide the search for the proof in the following ways.

 – Variables in goals were instantiated to particular points (i.e. constants) in such a way that the goal instances were true in the diagram.

- Clauses for which no such instantiation was possible were temporarily pruned from the tree.

- The nodes where such instantiations or prunings had been made were remembered as continuation nodes should the diagram ever be extended by construction and hence new points become candidates for instantiation.

Thus the diagram was used to guide the search by pruning some nodes from the tree and delaying the development of others to a second pass. We will call this technique *semantic checking*.

10.3 Constructions

If we discount loops then the resulting search tree is finite. This is because Gelernter's representation of Geometry does not use functions in any essential way: formulae like seg(A,B)=seg(D,C) can be regarded as fancy notation for a four parameter predicate. As noted in section 4.3.2, and section 16.2 function-free theories are decidable. In fact the Lush Resolution search trees for such theories are finite, so exploration will eventually cease, even if no proof has been found. At this stage a continuation node can be selected and used to suggest a single construction which would allow further development of the node.

The angle bisector example above can be proved without such a construction, as can many simple geometry theorems, so to see how the construction process works we will have to look at another example.

> *Statement* If the segment joining the midpoint
> of the diagonals of a trapezoid is extended
> to intersect a side, it bisects that side.

Diagram

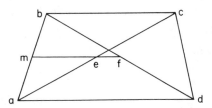

Given
→ quadrilateral(a,b,c,d)
→ seg(b,c) ∥ seg(a,d)
→ e midpt seg(a,c)

→ f midpt seg(b,d)
→ precedes(m,e,f)
→ precedes(a,m,b)

Required to Prove
seg(m,b) = seg(m,a) →

The meanings of the new functions and predicates introduced above is obvious except for 'precedes'. precedes(A,B,C) means that points A, B and C are collinear in that order.

We will consider a proof of this conjecture which uses the lemma that a line drawn through the midpoint, D, of the line, AB, to intersect AC at E, in triangle ABC, is parallel to the base, BC, iff E is the midpoint of AC. This is illustrated in figure 10-2.

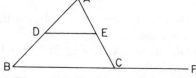

Figure 10-2: Key Lemma Illustration

A slightly extended version of the if part of this lemma is

is-triangle(A,B,C) &
precedes(A,D,B) & precedes(A,E,C) &
seg(D,A) = seg(D,B) & seg(E,A) = seg(E,C) &
collinear(B,C,F)
    → seg(D,E) || seg(B,F)   (iv)

The idea of the proof is to construct a line from c through f to intersect at at k. The resulting diagram is shown in figure 10-3. Point f is established to be the midpoint of ck, because triangles cfb and kfd are congruent. The lemma is used in the if direction, on triangle cak, to establish that ef is parallel to ak. Then the lemma is used in the only if direction, on triangle bad, to establish that m is the midpoint of ab. The hardest part is thinking of the construction of point k.

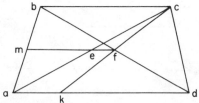

Figure 10-3: The Construction of a New Point

To see how the diagram can be used to suggest this construction consider the fragment of proof in figure 10-4, in which axiom (iv) is applied. This is the point of the proof at which ef is shown parallel to ak.

seg(e,f) ||seg(a,d) →

(iv)

$ is-triangle(X,a,Y) &
    precedes(X,e,a) & precedes(X,f,Y) &
    seg(e,X)=seg(e,a) & seg(f,X)=seg(f,Y) &
    collinear(a,Y,d) →

Figure 10-4: A Fragment of the Search Tree of the Second Example

During the first stage of the Geometry Machine's search, the goal clause marked $ in figure 10-4, was temporarily pruned, because there are no instantiations for X and Y *in the current diagram* which would make all the goals in $ simultaneously true. However, after the search tree had been exhausted it was resurrected as a continuation node.

But the goals of clause $ can be satisfied, and can only be satisfied, by the substitution {c/X, k/Y}. In fact, ruler and compass style manipulations, based on these goals, will construct k. Draw a circle, centre e, radius ea, and find its intersection with ec to bind X. Now extend cf to intersect ad at k. Check that fc and fk are equal.

The information that k is the intersection of cf and ad, i.e
       collinear(c,f,k) and collinear(a,d,k)

can now be added to the theorem hypotheses and the search can proceed from this continuation node. The proof of this example can now be found without further construction being necessary.

10.4 What is the Diagram?

What form does the Geometry Machine diagram take and how was it used?

Gelernter did not use a diagram written on a piece of paper and viewed, by the computer, through some electronic eye (although, if he had had the technology, this method could have been used instead). Rather, a cartesian representation was used, with each point mentioned in the theorem being assigned a pair of x y coordinates. Thus the diagram,

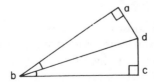

would be represented by some assignment like

$a \mapsto <1.5, \sqrt{3}/2>$
$b \mapsto <0, 0>$
$c \mapsto <\sqrt{3}, 0>$
$d \mapsto <\sqrt{3}, 1>$

chosen in such a way as to make the hypotheses of the theorem (i.e. all but the goal clause) true. The above assignment makes

$\rightarrow$ seg(d,b) bisects angle(a,b,c)
$\rightarrow$ seg(d,a) $\perp$ seg(b,a) and
$\rightarrow$ seg(d,c) $\perp$ seg(b,c)

all true. The assignment must be chosen by the human user of the Geometry Machine, since Gelernter did not provide a program to do it. The user is well advised to choose an assignment which minimizes the number of things which are accidently true.

The calculation procedures of real number arithmetic can now be used to assign real numbers to any variable free term and hence to assign a truth value to any variable free formula. For instance, tri(a,d,b)=tri(c,d,b) would be assigned the value t and tri(a,d,b)=tri(d,c,b) would be assigned the value f.

The diagram and these calculation procedures can be used to temporarily prune nodes out of the search tree as follows.

- Suppose G_1 & ... & $G_n \rightarrow$ is a newly derived goal clause containing the vector of variables, X. Suppose the vector of constants, a, contains all the constants mentioned in the statement of the theorem.

- The constants are assigned to the variables in the goal clause in all possible ways producing a series of variable free instances. Let G'_1 & ... & $G'_n \rightarrow$ be one of these instances.

- The calculation procedures and the assignment to a are used to test G'_1 & ... & G'_n to see whether it is true in the diagram.

- All instances failing the test are pruned.

- All instances passing the test become daughters of the original goal clause.

- If the original goal clause contains any variables it is remembered as a possible continuation node. Note that a variable free goal clause has only one variable free instance – itself. Thus the calculation procedures either reject or accept it. There is no need to remember it as a continuation node.

For instance, if the newly derived goal clause is

tri(a,d,X)≡tri(c,d,Y) →

then the variable free instances of this are:

tri(a,d,a)≡tri(c,d,a) →
tri(a,d,a)≡tri(c,d,b) →

.... etc (16 in all)

Only one of these passes the 'truth' test, tri(a,d,b)≡tri(c,d,b). So only this instance remains in the search tree.

Exercise 29: What are the variable free instances of seg(d,a)=seg(d,X) →? Which of them pass the truth test?

The other use of the diagram mentioned above was to help in the construction of new points. This is a little more difficult than the truth testing use. Given a conjunction of goals like:

is-triangle(X,a,Y) & precedes(X,e,a) & precedes(X,f,Y) &
seg(e,X)=seg(e,a) & seg(f,X)=seg(f,Y) & collinear(a,Y,d)

the diagram must suggest how new points may be constructed, which when substituted for some of the variables in the conjunction will make it true. In this case it should suggest the construction of point k, as the intersection of lines ad and cf and suggest the substitution {c/X, k/Y}.

The trick is to associate algebraic formula, e.g. equations, inequalities etc, with each of the propositions in the conjunction, and to use equation solving techniques to find an assignment of cartesian coordinates for the variables which will make the conjunction true. This assignment can then be used to recover the points corresponding to the coordinates: the points may already exist (e.g. c in the example) or may need to be defined as the intersection of two existing lines.

For instance, the formula associated with precedes(X,e,a) is the equation of a line passing through the coordinates assigned to e and a. The formula associated with seg(e,X)=seg(e,a) is the equation of a circle with centre the coordinates of e and radius the length of seg(e,a). Solving these two equations for their intersection will produce two coordinates for X, corresponding to the existing points, a and c. Only the substitution of c for X will make precedes(X,e,a) true.

The formulae associated with precedes(c,f,Y) and seg(f,c)=seg(f,Y) may now be used in similar fashion to calculate coordinates for Y. These correspond to no existing points, but define a point which lies on the non-parallel lines ad and cf, and so may be introduced as the intersection of these two lines.

The axiom which allows us to introduce new points

$$\sim \text{seg}(A,B) \ \| \ \text{seg}(C,D) \rightarrow \exists E \ \text{collinear}(A,B,E) \ \& \\ \text{collinear}(C,D,E)$$

is known implicitly to the Geometry Machine rather than being available to extend the search tree in the normal way. Including it as an explict axiom might cause problems since putting it into clausal form produces two non-Horn clauses. It is invoked only during the resurrection of continuation nodes in order to create new points and to assert the collinearity propositions on the right hand side of the axiom. In this case having checked that lines ad and cf are not parallel we can assert two new hypotheses about the new point, namely

$\rightarrow$ collinear(a,d,k) and
$\rightarrow$ collinear(c,f,k)

The procedure may be summarized as follows

– Given a continuation node labelled by a goal clause, $g_1 \ \& \ ... \ \& \ g_n \rightarrow$ containing the variables, X.
– Associate algebraic formulae with each of the goals, g_i and algebraic variables with each of the variables, X.
– Solve these formulae to assign coordinates to each of the variables, X. There may be several combinations of coordinates.
– Check that these coordinates satisfy all the g_is rejecting any combinations that do not.
– Recover points to associate with the variables by noticing that the assigned coordinates are already assigned to points or by creating new points as the intersection of two non-parallel lines.
– Assert collinearity propositions for the new points.
– Substitute the points for the associated variables in the goal clause creating a new instance for each combination of coordinates.
– Let each of these instances be a daughter of the original goal clause.

10.5 Can the Diagram be Generalized?

Is this use of the diagram just a technique useful in Geometry or can it be generalized and used in other domains?

It can be generalized. The key observation is to recognise that the job of the diagram above is to define a *model* of the hypotheses of the theorem and the axioms of Geometry – in the sense that we met in chapter 3.

Firstly, note that the diagram was an interpretation of the clauses which constitute the theorem statement, i.e. the axioms of geometry and the hypothesis and conclusion of the theorem. The universe of the interpretation was the set of all pairs of real numbers, e.g. $<\sqrt{3},1>$. Each constant in the theorem statement was assigned one of these pairs as its value, e.g. d $\mapsto$ $<\sqrt{3},1>$. Each function and predicate in the theorem statement was assigned an arithmetic calculation procedure.

Secondly, note that the assignment of coordinates to points was done so as to make this interpretation a model of the hypotheses of the theorem and the calculation procedures were chosen so as to ensure that it was a model of the axioms of Geometry.

To see how the same technique may be used in a different domain, consider the following example from arithmetic.

Let X ∤ Y mean that X does not divide Y exactly. We will try to prove the theorem that

If 5 does not divide a number then 30 does not divide it.

We will take as our axioms,

1. X ∤ Z → X.Y ∤ Z
2. Y ∤ Z → X.Y ∤ Z
3. → 30=2.(3.5)

and the equality axioms, which we will build into the paramodulation rule.

The theorem is formalized as:

hypothesis: → 5 ∤ a
        and
conclusion: 30 ∤ a →

The search tree for this example, taking the conclusion as top clause, is given in figure 10-5.

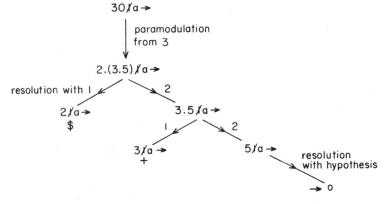

Figure 10-5: Search Tree for Not Divides Example

This tree, though exhausting the possibilities, is nice and small. However, if we had had some additional axioms for χ, e.g.

3. $Z \mid X$ & $Z \chi Y \rightarrow X \chi Y$,

then the goal clauses marked $ and + may have been developed indefinitely and fruitlessly. It is worth looking at ways in which they could have been recognised as hopeless and pruned from the tree.

Let us use the calculation procedures of arithmetic to build an interpretation which is a model of the axioms and hypotheses of the theorem, but in which the goals of $ or + are false. One such interpretation, arith2, can be defined by letting ., χ and = have their normal meaning, but giving 'a' the value 2. In arith2, 5 χ a, for instance, will be true, and so will 30 χ a, but 2 χ a will be false. Thus arith2 rejects the goal clause $.

However, arith2 will not reject +, since it makes the goal, 3 χ a, true. To reject + we will need an interpretation which makes 3 χ a false. Consider, for instance, the interpretation, arith3, in which ., χ and = have their normal meanings, but 'a' has value 3. This has the desired effect. It rejects +. However, it does not reject $. Thus to reject both $ and + we will need to use both arith2 and arith3 in concert.

This example has two messages:

 – The use of the diagram by the Geometry Machine to prune some goal clauses from the search tree can be generalized to other domains: the diagram being generalized to an interpretation of the clauses and a model of the axioms and hypotheses.

 – It can also be generalized in another way, namely several interpretations can be used in concert, any one having the right of veto.

Exercise 30: Can you find an interpretation which is a model of the axioms and hypothesis, but which rejects both $ and + ?

Exercise 31: Consider the search tree for the group theory example in figure 7-1, chapter 7. Can you design a model of the axioms and hypothesis of the theorem which would reject the goal clause a=b → ?

10.6 The Trouble with Non-Horn Clauses

So far we have argued by example (an unsafe method). Before we proceed to a theoretical analysis let us look at one more example – a counterexample.

Consider another funny arithmetic example. The proposition, is-prod-primes(X), means X is a product of prime numbers, is-prime(X) means X is a prime number and is-power-3(X) means X is a power of 3. As axioms we will have:

1) is-prime(X) → is-prod-primes(X)
   and
2) is-power-3(X) → is-prod-primes(X)

i.e. a number is a product of primes if it is a prime or a power of 3.

We will now try to prove this as a theorem. Negated, this becomes the hypothesis,

hypothesis: → is-prime(a) v is-power-3(a)

and the conclusion,
   conclusion: is-prod-primes(a) →.

Notice that the hypothesis is a non-Horn clause.

If we take the conclusion as the top clause and apply Lush Resolution we get the complete search tree shown in figure 10-6.

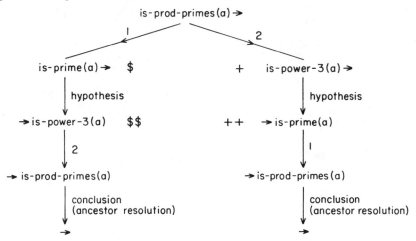

Figure 10-6: A non-Horn clause search tree

Notice also the non-Horn clause features of this search tree:
   – that the clauses labelling some nodes are not goal clauses, even though the top clause was;
   – that it is necessary to use ancestor resolution with the top clause.

I am going to present some interpretations which, if allowed right of veto, would prune all proofs from this search tree – that is, the use of semantic checking will make the theorem proving process incomplete!

As usual we will let the predicates: is-prime, is-power-3 and is-prod-primes, take their obvious meanings on the natural numbers, 0, 1, 2, 3, ...etc. Thus is-prime(2), is-power-3(9), is-prod-primes(2) and is-prod-primes(9) will be true and is-prime(9) and is-power-3(2) will be false. So any interpretations we consider will be models of the axioms, 1 and 2. The various interpretations will differ only in the meaning assigned to the constant a. In order that they will be models of the hypothesis, we must ensure that 'a' behaves like a prime or a power of 3.

In the first interpretation, num2, we will give 'a' the value 2. Thus, is-power-3(a) will be false so num2 will reject the goal clause +.

In the second interpretation, num9, we will give 'a' the value 9. Thus, is-prime(a) will be false, so num9 will reject the goal clause $.

If both num2 and num9 are used in concert with right of veto then the search tree will be truncated after the first level and both the proofs will be lost.

The solution might appear to be to drop the right of veto, and insist that both interpretations agree, or to use only one interpretation. However, neither of these possible solutions is good enough. We can find a single interpretation which will prune both branches.

Consider the interpretation, num3, in which 'a' is given the value 3. This will reject both clauses $$ and ++, provided we extend the rejection criteria in the obvious way to non-goal clauses. Clause ++ is → is-prime(a). This is equivalent to ∼ is-prime(a) →, i.e it behaves like a goal clause with goal ∼ is-prime(a). But is-prime(a) is true in num3, so ∼ is-prime(a) is false and num3 will reject ++. Similarly num3 will reject $$.

What went wrong? Why did our nice semantic based guidance technique go sour? Could it be something to do with the fact that this example is a non-Horn clause? To find out we will have to take the theoretical excursion promised at the beginning of this section.

10.7 The Theoretical Underpinning for Semantic Checking

Let us start by considering, and tidying up, the grounds on which a clause can pass the 'truth test' set by an interpretation.

On page 140 above we said that an instance of a goal clause

$$G'_1 \& \ldots \& G'_1 \to,$$

would pass the truth test if and only if all the G'_is were true in the interpretation. This goal clause instance is equivalent to:

$$\sim G'_1 v \ldots v \sim G'_n$$

so if all of the G'_is are true then the whole clause is false. Any instance of the goal clause which is *false* in the interpretation will be accepted and any true instance will be rejected.

Let us extend these grounds to any variable free clause and say that

> *Definition 1:* An interpretation *accepts* a variable free clause if and only if it is false in the interpretation.

One effect of semantic checking is to ensure that any clause, which is true in a model of the axioms and the hypothesis of the conjecture, is pruned from the tree. We will show that this cannot lead to incompleteness by proving the following theorem.

> *Theorem 2:* If input U {top} is a minimally unsatisfiable set of Horn clauses then there is a derivation of the empty clause from them in which 'top' is the top clause and no clause in the derivation is true in any model of 'input'.

> *Proof:* The results of section 16.5 established that Lush Resolution is complete for sets of Horn clauses, so there is a derivation in which top is the top clause and each clause in the derivation is derived by resolution with a clause from input. We will show that this derivation has the required property.

> Let M be a model of input. Since input U {top} is unsatisfiable, top is not true in M. Since the empty clause is not true in any model it is not true in M. We will show that if a resolvant is not true in M then one of its parents is not true in M, and hence by induction on the structure of the derivation, all the clauses in it are not true in M.

> To simplify matters consider only a binary resolution between:

> C' v P' and C" v ~P"

> producing the resolvant

> (C' v C")φ

> where φ is the most general unifier of P' and P". Suppose that C' v P' and C" v ~P" are true in M and that (C' v C")φ is not.

> Then there is a variable free instance of the resolvant (C' v C")φ which is false in M. Hence both C' φ and C" φ are false in M. Since the original clauses are true then all their instances are true, so P'φ and ~P"φ must both be true. But P'φ and P"φ are identical, so this is a contradiction. Therefore in order to produce a resolvant which is not true one of the its parents must itself be not true. This argument is easily extended to full resolution.

The bottom clause of the derivation is the empty clause, which is not true in M.

Assume, as an induction hypothesis, that the clause nth from the bottom in the derivation is not true in M. We will show that the n+1st from the bottom is also not true in M and hence, by induction that none of the clauses in the derivation are true in M.

The nth clause was derived by resolution from the n+1st clause and a clause from input. By assumption the input clause is true in M and by the induction hypothesis the nth clause is not true in M. Hence, by the above argument the n+1st clause is not true in M. Therefore, none of the clauses in the derivation are true in M.   QED

This theorem partially justifies the pruning of semantic checking. It shows that the clauses which were completely pruned, the ones true in some model of the input clauses, could not have led to any proofs and were rightly ignored. If we add to Lush Resolution an instantiation rule which instantiates variables with elements from the universe and show that completeness is preserved if after this rule is applied no resolutions are necessary with the parent clause then the theorem completely justifies semantic checking.

However, the theorem breaks down for non-Horn clauses at the induction step. Because ancestor resolution is allowed, the nth clause may have been derived from the n+1st clause and an ancestor, say the n+mth clause from the bottom. It could be that the n+1st clause is true, but that the n+mth clause is not true and that the nth clause inherits its non-truth from that. Thus truth can be introduced into the derivation only to be eliminated by ancestor resolution.

Exercise 32: Verify that that is what happens with the non-Horn clause example of section 10.6.

10.8 Summary

Thus the situation can be summarized as follows. For Horn clauses:

1. a model of the input clauses (axioms plus conjecture hypotheses) can be used to prune the search space of true clauses without loss of completeness;

2. several models can be used, with each having the right of veto.

3. in the case of Geometry the model can be thought of as a diagram, but semantic checking is applicable to any domain;

4. in Geometry, a blocked search tree can be used in conjunction with the diagram to suggest the construction of new points and lines.

It may be possible to adapt semantic checking to non-Horn clauses. For instance, by insisting that if a true clause is introduced attempts are made to ancestor resolve it or its descendants with a non-true ancestor. To the best of my knowledge no one has attempted this.

Further Reading Suggestions

[Gelernter 63, Gelernter et al 63] are the best introduction to the Geometry Machine. [Gilmore 70] is an analysis and rational reconstruction of Gelernter's techniques.

11

The Productive Use of Failure

☐ This chapter shows how evidence from the failure of one proof
method, namely Symbolic Evaluation, can be used to guide other
methods, namely Induction and Generalization.
☐ Section 11.1 defines LISP as a Predicate Logic theory.
☐ Section 11.2 describes the Symbolic Evaluation method of proof.
☐ Section 11.3 describes the Induction method.
☐ Section 11.4 describes the Generalization method.
☐ Section 11.5 explains how the Boyer/Moore technique can be
applied to Peano Arithmetic.

The uniform search strategies we have considered so far, e.g. depth first
search, have involved giving up in one part of the search tree and continuing
in another part, without any effort to investigate why the earlier attempt
failed. This contrasts with the behaviour of human mathematicians, whose
later attempts may be heavily influenced by their earlier failures (cf, for
instance, the accounts of [Waerden 71, Lakatos 76] mentioned in section
1.3.2).

By observing their own proofs and those of others, Bob Boyer and J
Moore noticed that this was especially true when proving theorems about
recursive functions. Two basic proof methods are available: a simple one in
which rewrite rules derived from the recursive definitions are used to
symbolically evaluate the theorem to be proved and a more complex one in
which induction is used to divide the theorem into two simpler theorems.
Boyer and Moore observed that the failure of the first method could often
suggest what induction schema to use and what variable to induct on, in the
second method.

They tested these ideas by building a theorem prover for proving the
properties of computer programs, using Symbolic Evaluation and mathematical induction. The theorem prover, universally known as the
Boyer/Moore Theorem Prover, [Boyer and Moore 73], has been under
continuous development since 1971. A more recent version of the system is
described in [Boyer & Moore 79]. This chapter is a rational reconstruction
of the program, paying particular attention to how the failure of Symbolic
Evaluation suggests how to use Induction. It is based mainly on early
versions of the program, since these are simpler to explain, while containing

the key ideas.

The domain of the Boyer/Moore theorem prover was a particular programming language, LISP, [McCarthy et al 62]. We can regard it as a mathematical theory of the properties of lists in which the terms of the theory also have an an interpretation as computer programs. Their techniques will work on any theory with recursive definitions and induction, e.g. Peano arithmetic. However, LISP has a richer collection of induction schemata than arithmetic, and so provides a better illustration of the power of the method.

11.1 The Formal Theory of LISP

As defined in section 4.3.3 a list is an ordered set of objects, represented as a series separated by commas and delimited by angle brackets, e.g.

<this, is, a, list>.

The elements of the list can be constants or other lists, e.g.

<this, is, <a, nested, list>>.

The empty list is represented by the constant nil.

The first list manipulation function we will consider is the constructor function, *cons*. cons is a binary function, whose second parameter must be a list. Its effect is to join its first parameter onto the front of this list, e.g.

cons(this, <is, a, list>) = <this, is, a, list>
and
cons(<first, element>, <of, a, list>) = <<first,element>, of, a, list>

cons plays a similar role, in LISP, to that played by s (the successor function) in Peano arithmetic. Just as s was used to generate all the natural numbers from the single constant 0, so cons can be used to generate all lists from a small collection of constants, e.g. <this, is, a, list> can be represented by

cons(this,cons(is,cons(a,cons(list,nil)))).

As with the successor function, this is a theoretically useful representation, but is rather cumbersome in practice, so we will maintain the fiction that the angle bracket representation is only an abbreviation for the underlying cons representation. This is just like the situation in Peano arithmetic, where 3 was regarded as an abbreviation of s(s(s(0))).

We will take this process a stage further by limiting the constants we will allow to one, nil. All other constants will be regarded as an abbreviation for some combination of cons and nil. For instance, the natural numbers, 0, 1, 2, 3, ... etc, will be represented by the lists, nil, <nil>, <nil,nil>, <nil,nil,nil>, ... etc. We will find it convenient to have, as terms within the theory, two symbols which behave rather like truth values. We will also use the symbols *tt* and *ff* to denote these, in order to distinguish them from *t* and *f*. We will adopt the convention that *tt* and *ff* are abbreviations for the lists cons(nil,nil) and nil, respectively.

Exercise 33: Translate the lists, <nil> and <nil,nil,nil> into combinations of cons and nil.

In Peano arithmetic we had an axiom ~ 0 = s(X) to ensure that no amount of applying the successor function brought us back to 0. In LISP we will want a similar axiom to ensure that no amount of consing can get us back to the empty list.

1) ~ nil = cons(X,Y)

As duals of the constructor function, cons, we will have the destructor functions, *car* and *cdr*. The car of a list is the first element, e.g.

car(<this, is, a, list>) = this.

The cdr of a list is the list minus its first element, e.g.

cdr(<this, is, a, list>) = <is, a, list>.

The relationship between car, cdr and cons, can be summarized in the axioms.

2) car(cons(X,Y)) = X
and
3) cdr(cons(X,Y)) = Y

In order to define new functions by cases we will need a conditional function, *cond*. cond is a ternary function. If the first parameter of cond is ff then it equals its third parameter, otherwise it equals its second. This can be expressed by the two axioms.

4) cond(ff,U,V) = V
and
5) cond(cons(X,Y),U,V) = U

Since LISP contains an equality predicate, *equal*, between lists we will have to represent this in our theory. Following our convention that all LISP programs are to be represented as terms, we must represent equal as a function. This distinguishes it from the predicate, =, which we have been using above. Indeed we have axioms which involve them both, like

6) equal$(X,X) = tt$

Finally, we are allowed to introduce as many new functions as we like, provided they are accompanied by new axioms which serve to define them either explicitly or by recursion. For instance, we may define a function, append, by recursion, as follows.

append$(X,Y) = $ cond$(X,$
                        cons$($car(X), append$($cdr(X), Y$))$,
                        $Y)$

append is a function which takes two lists and joins them, e.g.

append$(<$this, is$>, <$a, list$>) = <$this, is, a, list$>$

The idea of the definition is that if the first parameter is the empty list then we return the second parameter, otherwise we take off its first element, recursively append this foreshortened list to the second parameter and then cons back the first element. Note that the use of cond enables us to dispense with the need for multiple equations when making recursive definitions. This will have consequences for the way in which the failure of Symbolic Evaluation influences the subsequent induction process.

11.2 Symbolic Evaluation

A typical conjecture which we might want to prove in list theory is that appending a nil onto a list has no effect, i.e.

equal$($append$($nil$,X), X) = tt$ (i)

Taking the universal closure of this, negating it and putting it in clausal form gives:

$equal(append(nil,x), x) = tt \rightarrow$

The empty clause can be derived from this using Symbolic Evaluation alone, i.e. by rewriting the expression in italics into tt, yielding:

$tt = tt \rightarrow$

and then, using the reflexive law, into $t \rightarrow$ which is equivalent to the empty clause.

All the theorems we will prove in this chapter will normalize into the form:

exp $= tt \rightarrow$

and their proofs will only differ in the manner in which exp is Symbolically

Evaluated to *tt*, so this is the only part we will mention. In the case of conjecture (i) this means considering how

equal(append(nil,x), x)

can be Symbolically Evaluated to *tt*.

In this particular proof we will rewrite the expression using the rules:

append(X,Y) => cond(X,
                cons(car(X), append(cdr(X), Y)),
                Y)

cond(ff,U,V) => V
equal(X,X) => *tt*

in that order.

Note that, as in the case of arithmetic evaluation in section 9.2.3, the first rewrite rule is based on the recursive defining equation of append. The remaining rules are based on the non-recursive definitions of cond and equal. In general, we will use only rewrite rules based on the recursive definitions of new functions like append and those based on the axioms 1) to 6).

Thus the proof is:
equal*(append(nil,x), x)* = equal*(cond(nil,*
                      cons(car(x), append(cdr(nil),x)),
                      x),

                  x)

        = *equal(x,x)*

        = *tt*

where the subterm being rewritten at each stage has been emphasised in italic type.

But Symbolic Evaluation does not always work so well. The rewrite rule set defined by the axioms 1– 6 and recursive definitions is not terminating. For instance, a non-terminating branch can be obtained by applying the definition of append, repeatedly, to the term, append(a,b).

   append(a,b) = cond(a,
                    cons(car(a), *append(cdr(a),b)),*
                    b)

        = cond(a,

cons(car(a),
    cons(cdr(a),
        cons(car(cdr(a)),
            append(cdr(cdr(a)),b)),
        b),
    b)

= .... (and so on)

Note that each successive appearance of append has a more deeply nested first parameter: a, cdr(a), cdr(cdr(a)), ... etc. This reflects the definition of append, where the first element of the first parameter was repeatedly removed until the empty list was revealed. But a Skolem constant like 'a' represents an arbitrary list – there is no knowing how long it is. Consequently, the process will never terminate. The loss of termination is due to our use of cond to define recursive functions in one equation. If we had used two equations, as in section 9.2.3, e.g.

append(nil,Y) => Y
append(cons(X1,X),Y) => cons(X1,append(X,Y))

then Symbolic Evaluation would terminate with a non-specific list.

If the result of rewriting an expression is that car or cdr is applied to one of its parameters then two cases arise:

– either the dominant function of the parameter is cons, in which case the car or cdr can be eliminated by axiom 2) or 3)
– or the dominant function is something else, in which case the car or cdr cannot be eliminated.

This second case is the major cause of non-termination. We will call such a parameter an *ugly expression*. Quite often these ugly expressions are Skolem constants, as the example above illustrates. If a rewriting threatens to produce an ugly expression then the rewriting is prohibited and Symbolic Evaluation is terminated. We will see that the ugly expressions not only tell us when to stop Symbolic Evaluation, but also provide a strong clue about what to do next.

11.3 The Method of Induction

What to do next is to try the other available method, Induction. But here we are faced with a choice. There are several induction schemata available, and each of them can be used to induct on a different parameter.

Here are a selection of some of the induction schemata available in list theory.

P(nil) & ∀X ∀X1[P(X) → P(cons(X1,X))] → ∀X P(X)

P(nil) & ∀X1 P(cons(X1,nil))
        & ∀X ∀X1 ∀X2[P(X) → P(cons(X2,cons(X1,X)))] → ∀X P(X)

P(nil) & ∀X ∀X1[P(X1) & P(X) → P(cons(X1,X))] → ∀X P(X)

In the first one induction is on the cdr of the list; in the second it is on the car and in the third it is on both.

In each of these the role of X may be taken by any Skolem constant in the formulae being proved, e.g. either a or b above, making 6 possibilities altogether. And things do not stop there, as other induction schemata may be used, in which, for instance, the induction hypothesis covers all sublists of X.

This is where the failed attempt at Symbolic Evaluation comes in. Remember that in the example above the first rewriting produced append(cdr(a),b). If we had assumptions about this formula we might have been able to do something more than get involved in an infinite regress. But such an assumption is precisely what an induction hypothesis offers us, provided we choose the right one. In this case since only the Skolem constant 'a' is being effected, and since it is its cdr that is involved then the first induction schema above is strongly suggested with 'a' substituted for X.

Let us see how this might work in the case of a real theorem proving attempt. Consider the conjecture, that append is an associative function, represented by:

equal(append(A,append(B,C)), append(append(A,B),C)) = tt

If we try to prove this by Symbolic Evaluation, i.e. by rewriting

equal(append(a,append(b,c)), append(append(a,b),c))

into tt, then we will generate four branches – one for each occurrence of append. Each of these will be terminated by the threatened application of a destructor function, car or cdr, to an ugly expression.

Looking back at our attempt to Symbolically Evaluate append(a,b) we can see that in each of the four cases the ugly expression will be the parameter of an occurrence of cdr dominated by the function, append. The ugly expressions involved will be two occurrences of a, one of b and one of append(a,b). This is diagrammed in figure 11-1

The cdrs all suggest the first induction schema above, i.e. induction on the

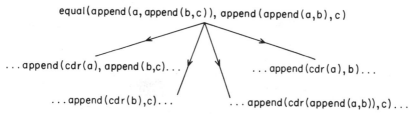

Figure 11-1: Potential Infinite Regresses in Rewriting of append Example

cdr of a list. But what list? The candidates are a and b, and as we will see below, it is possible that append(a,b) could be generalized into a constant and become a candidate. Note that both occurrences of a give rise to candidates, whereas only one occurrence of b does. If b were chosen as induction variable then the fact that only one of its occurrences would be unpacked by Symbolic Evaluation would cause trouble during the induction step. An induction with b as variable is called a *flawed induction* and one with a as variable is called an *unflawed induction*. The Boyer/Moore theorem prover would choose a as the induction variable.

Thus the induction basis is to prove.

equal(append(nil,append(b,c)), append(append(nil,b),c)) = *tt* →

which yields easily to Symbolic Evaluation (Note how similar it is to the example we did above). The induction step is to assume the hypothesis

→ equal(append(a,append(B,C), append(append(a,B),C)) = *tt*

and then prove

equal*(append(cons(a1,a),append(b,c)),*
    append*(append(cons(a1,a),b),c)) = tt* →

The left hand side of this can be rewritten to

equal(cons(a1,*append(a,append(b,c))),*
    cons(a1,append(append(a,b),c)))

Now the induction hypothesis can be applied, as if it were a rewrite rule, to replace the left hand side double append and get

equal*(cons(a1,append(append(a,b),c)),*
    cons(a1,append(append(a,b),c)))

which is trivially rewritten to *tt*.

The step where the induction hypothesis was used was not the same kind of rewriting step as the others because the induction hypothesis had the form equal(lhs,rhs) = *tt*, rather than lhs = rhs. Boyer and Moore call this an application of *Fertilization*.

In summary then, the Boyer/Moore technique is

1. Try to prove the theorem by Symbolic Evaluation.
2. Trap any infinite branches by looking out for a destructor function applied to an ugly expression parameter of the subterm about to be rewritten.
3. If the attempt at Symbolic Evaluation fails try Induction.
4. Use the trapped destructor function(s) to suggest the induction schema to use and the trapped ugly expression(s) to suggest what should be substituted for the induction variable, prefering those that lead to unflawed inductions.
5. Try to prove the induction basis and conclusion. Use Fertilization during the proof of the induction step.

11.4 Generalizing the Theorem to be Proved

It is often the case in proving theorems by Induction that the induction hypothesis is not strong enough to support the induction conclusion. When this happens it may be that we should try a different induction schema or it may be that the theorem should first be strengthened by generalizing it – a paradox of proofs by Induction is that it sometimes easier to prove a strong theorem than a weak one, simply because the induction hypothesis is also stronger. This is one of the major points made in [Waerden].

This is the case with the theorem

equal(append(*rev(A)*, append(B,C)),
        append(append(*rev(A)*,B),C)) = *tt*

where rev is a function which returns a list in the reverse order, e.g.

rev(<this, is, a, list>) = <list, a, is, this>

The induction hypothesis is just not strong enough to do its job and the Boyer/Moore technique above, fails.

However, we have already seen that a stronger version of the theorem can be easily proved, namely the theorem obtained by replacing rev(A) above by a variable, say D.

equal(append(*D*,append(B,C)),
        append(append(*D*,B),C)) = *tt*

This replacement of a term by a variable is called *Generalization*. It can be done before Induction is attempted to improve the chances of the attempt succeeding. An obvious procedure suggests itself, namely to look for two or more identical, complex subterms (i.e. not variables or constants) in a formula and replace them with new variable. Thus the subterm, append(b,c), in the theorem above would not be replaced because it only occurs once and b would not be replaced because it is a constant.

A further refinement (due to Aubin [Aubin 75]) is to insist that the term to be replaced should be one of the trapped ugly expressions. The reasoning behind this is that for Generalization to achieve anything it should make available an induction variable candidate that was not available before. Induction variable candidates must all show up as trapped ugly expressions, but those that are complex terms cannot be used as they stand. Generalizing them to variables would make them available. This works in the example above, as rev(A) is trapped twice as an ugly expression. Thus the failure of Symbolic Evaluation can not only suggest how to apply Induction, but also how to Generalize the conjecture to be proved.

The danger in Generalization is to over-generalize, i.e. to generalize a theorem into a non-theorem. An example of this is provided by the theorem,

$$\text{equal}(\text{length}(\mathit{length(X)}), \mathit{length(X)}) = tt$$

where length is a function from lists to numbers, which in our formal theory are played by lists of nils. Thus

```
length(length(<this, is, a, list>))
        equals length(<nil, nil, nil, nil>)
        equals <nil, nil, nil, nil>
        equals length(<this, is, a, list>)
```

This theorem contains two occurrences of the complex term, length(X), one of which will be trapped as an ugly expression during the Symbolic Evaluation of length(length(X)). If these are replaced by a new variable, say Y, we get the formula

$$\text{equal}(\text{length}(Y), Y) = tt \qquad \qquad \text{(ii)}$$

which is no longer a theorem! In fact (ii) is only true when Y is a number, i.e. a list of nils.

Boyer and Moore provided a partial answer to this problem of over-generalization by having their theorem prover attempt to invent an additional hypothesis about the newly introduced constant which would retain the truth of the theorem. In this case the theorem prover would augment (ii) with an additional hypothesis to the Generalization:

is - number(Y) = tt → equal(length(Y), Y) = tt

where is - number(Y) means Y is a list of nils. This formula is a theorem.

For this additional hypothesis to be meaningful any functions in it would have to be defined. In this case additional axioms would have to be asserted to define is - number. The theorem prover was able to do this in many cases, merely by examining the replaced subterm. In this case by examining the term length(Y) to see what kind of list it might be.

In inventing these axioms the Boyer/Moore Theorem prover was nearly always successful, but sometimes it would over-generalize and the additional hypothesis would admit too many lists. This is why the solution was only partial. However, if the Generalized theorem could be proved, then the original theorem would follow from it.

Another way to avoid over-generalization (again due to Aubin), is to further refine the heuristic for suggesting Generalization candidates. Instead of just noting complex trapped ugly expressions as Generalization candidates, we could note which occurrences of the expression are trapped. When Symbolically Evaluating

equal(length(length(X)), length(X)) = tt

the first occurrence of length(X) is trapped, but the second is not. This observation can be used to block the Generalization of length(X), since the induction candidate should appear on both sides of the equality to be effective.

11.5 Applications to Arithmetic

The Boyer/Moore technique is applicable to any theory in which functions are defined recursively and theorems proved by Induction. In this section we show how this can be done to the most well known such theory, Peano Arithmetic.

In arithmetic, the role of the recursive data structure, lists, is played by the natural numbers, 0, s(0), s(s(0)), ... etc. The successor function, s, plays the part of the constructor function, cons, and the predecessor function, p, plays the part of the destructors, car and cdr, where

p(s(X)) = X   and   p(0) = 0

We need to introduce a new function, cases, to play the part of cond, i.e.

cases(0,Y,Z) = Y   and
cases(s(X),Y,Z) = Z

which returns its second or third parameter, according to whether its first parameter is 0 or not. These equations defining p and cases must be used by Symbolic Evaluation, as left to right rewrite rules.

To use the Boyer/Moore technique as it stands, we must modify the recursive definitions of the arithmetic functions to be single equations using the cases function, e.g.

$$X+Y = cases(Y, X, s(X+p(Y)))$$

Compare this with the definition of + given in section 4.2.3. We can use this definition as a rewrite rule, left to right, during Symbolic Evaluation.

We are now in a position to consider the proof of the associativity of +.

$$(X+Y)+Z = X+(Y+Z)$$

To be proved by the Boyer/Moore technique this conjecture must be represented as the goal clause.

$$equal((x+y)+z, x+(y+z)) = tt \longrightarrow$$

Symbolic Evaluation of this goal clause, using the definition of +, yields the four terminated branches diagrammed in figure 11-2.

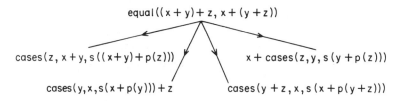

Figure 11-2: Potential Infinite Regresses in Rewriting of + Example

In each of the four cases the trapped ugly expressions are parameters of the predecessor function, p, dominated by the + function. The four ugly expressions are: z, y, y+z and z. The complex term, y+z, would be a candidate for Generalization, if it appeared more than once, but it is not yet a candidate for the induction variable. Of the remaining candidates, y and z, z wins because it leads to an unflawed induction.

Since the trapped occurrences of z both appear as parameters of p then the suggested induction scheme is the standard one,

$$Q(0) \& \forall X \{Q(X) \rightarrow Q(s(X))\} \rightarrow \forall X Q(X)$$

Applying this to our theorem generates the two subgoals:

Basis
$$equal((x+y)+0, x+(y+0)) = tt \rightarrow$$
and

Step

$\rightarrow$ equal((X+Y)+z, X+(Y+z)) = *tt*
equal((x+y)+s(z), x+(y+s(z))) = *tt* $\rightarrow$

The basis rapidly yields to Symbolic Evaluation. The step can be proved by Symbolic Evaluation plus Fertilization with the induction hypothesis. The proof is strongly parallel to that of the associativity of append, given earlier.

Exercise 34: Describe the proof of the step above, as it might be found by the Boyer/Moore technique.

Exercise 35: Describe how the Boyer/Moore technique would prove the conjecture, (X+s(Y))+Z = X+(s(Y)+Z).

Thus the Boyer/Moore technique can be applied to theories other than LISP.

11.6 Summary

The Boyer/Moore Theorem Prover has available a variety of methods of proving theorems: Symbolic Evaluation; Induction; Fertilization and Generalization. It understands something of the relation between these methods. It knows when to apply each one and when the failure of one method can help in the application of another. In short, it can reason about the different methods available. We will take this idea a step further in the next chapter.

Further Reading Suggestions

[Boyer and Moore 73] is a short and readable introduction to the Boyer/Moore theorem prover, although it is a little out of date. [Boyer & Moore 79] is a longer and more up to date account. [Moore 74] gives a detailed account of the implementation of the original program.

12
Formalizing Control Information

☐ In this chapter we see how to represent control information as a Predicate Logic theory and use inference in this theory to guide search.

☐ Section 12.1 introduces some control information that can be used to guide the search for a solution to an equation.

☐ Section 12.2 describes various equation solving solution methods: Isolation, Collection and Attraction.

☐ Section 12.3 defines the Meta-Theory of Algebra and shows how to represent control information with it and how to use this information to guide the search for a solution.

When a human mathematician has built up expertise, in a particular area of mathematics, he has at his disposal a variety of problem solving methods. He is able to bring these methods to bear on a problem, choosing an appropriate method for the goal he is trying to achieve and the situation in which he finds himself.

We saw, in the last chapter, some examples of such methods: Symbolic Evaluation, Induction, Generalization, etc. In this chapter we take this investigation a stage further and ask how such methods should be organised, so that they may be brought smoothly into operation when the occasion demands and be combined together to tackle complex problems. The answer we offer is: that a process of reasoning should take place about the problem to be solved and the methods available to solve it; and that the techniques we have met in earlier chapters can be used to represent this reasoning process. It is precisely such reasoning processes which are advocated by Polya in the books mentioned in section 1.3.2.

The richest examples of multiple problem solving methods can be found in some of the older branches of mathematics, where experience has accumulated over several centuries. Consider, for instance, the problem of solving symbolic, transcendental equations, like

$$\log_e (x+1) + \log_e (x-1) = c$$

There are a variety of well known methods to deal with subclasses of equations, e.g. Gaussian elimination for sets of simultaneous linear equations, completing the square for quadratic equations and some unnamed, but well known, methods for trigonometric equations.

But what about those equations, like the one above, for which there are no established methods? Most mathematicians would find the above example quite easy – they would not indulge in search, but would proceed directly to a solution. This behaviour cannot be accounted for using the theorem proving techniques we have considered so far. This equation and the axioms of algebra, will create a huge search tree. Searching this tree in an unguided fashion will lead to a combinatorial explosion. So how do they do it? Could it be that they have available a variety of 'methods' which are not publically available in the textbooks of algebra, but are picked up from examples – the worked examples from textbooks and the successful personal solutions of the past?

12.1 Reading Between the Lines

One way to find out is to examine some worked examples or successful personal solutions and see if there is any pattern there, i.e. to analyse the solutions. Consider the following solution of the log example above.

$$\log_e (x+1) + \log_e (x-1) = c$$

...1

$$\log_e (x+1).(x-1) = c$$

...2

$$\log_e x^2 - 1 = c$$

...3

$$x^2 - 1 = e^c$$

...4

$$x^2 = e^c + 1$$

...5

$$x = \pm \sqrt{e^c + 1}$$

In order that we can read between the lines, I have labelled the spaces between them with numbers.

The most revealing part of this solution is at the bottom, so we will start our analysis there. Consider the last three steps, labelled 3, 4 and 5. There is a pattern here. At each step the dominant function on the left hand side is removed and its inverse becomes the dominant function on the right hand side: thus log is replaced with exponentiation; minus with plus and square with square root. The effect is to gradually decrease the depth of the unknown x, until it is totally isolated on the left hand side, with the solution on the right. This is one of the most common 'unwritten' methods of equation solving. It will work with equations composed of any sort of functions, provided each function has an inverse. We will call it *Isolation*.

Isolation will only work with equations containing a single occurrence of the unknown. If there are several occurrences, as in the first and second lines of the solution above, then one of the occurrences can be isolated, but the other occurrences will be moved to the right hand side, polluting the potential solution with their presence. Armed with this observation we can now make sense of step 2.

In step 2 the number of occurrences of x is reduced from 2 to 1, using the 'difference of two squares' identity,

$$(U+V).(U-V) = U^2 - V^2.$$

This has the effect of making the Isolation method applicable. We will assume that there is a method available for reducing the number of occurrences of an unknown. We will call it *Collection*, since it collects together occurrences of unknowns. Collection can often be seen at work in equation solving. It does not have the same guarantee of success as Isolation, but an identity can often be found to collect together two occurrences, especially if they are close together.

This last observation helps us to understand step 1. The number of occurrences of x is not reduced, but they are brought closer together, thus making it more likely that they can be collected. Again we will assume that there is a method for doing this and we will call it *Attraction*.

Exercise 36: Consider the equation

$$(2^{x^2})^{x^3} = 2.$$

Can you solve this using the methods of Isolation, Collection and Attraction?

12.2 Equation Solving Methods

How can Isolation, Collection and Attraction be represented as computer programs?

This becomes clearer when we see what algebraic formulae they will need to apply in order to achieve their stated purpose.

12.2.1 Isolation

For instance, Isolation applies the following double implications in the solution above.

$$\log_U V = W \longleftrightarrow V = U^W$$

$$U-V = W \longleftrightarrow U = W+V$$

$$U^2 = V \longleftrightarrow [U = \sqrt{V} \text{ v } U = -\sqrt{V}]$$

These can be organised as a set of rewrite rules . Isolation will apply them successively until the unknown is totally isolated, when no more rules will apply.

Notice that the rules above all have a similar structure:

- the left hand side is an equation between a term and a variable;

- the term consists of a function with variable parameters;

- the right hand side is either an equation between one of these variables and a term or is a disjunction of such equations;

- this term contains the inverse of the left hand side function.

This structure can be summarized as:

$$F(U_1,...,U_i,...U_n) = V \longleftrightarrow V_j \text{ } V_i = T_i^j(U_1,...,V,...U_n)$$

where U_i is the variable being isolated

It is not surprising that all the rules should have this structure as it is just what is required to achieve the Isolation effect of stripping away the dominant function of the left hand side until the unknown is isolated. It does mean that the Isolation method can be very selective about how it applies the rules under its care.

- It need only attempt to rewrite complete equations

- and only then if the distinguished variable U_i is matched to a term containing the unknown it is trying to isolate.

We will call these conditions the *selection criteria* of Isolation.

The last criterion can be met if each rule is stored with the parameter position it isolates and the position of the unknown occurrence is calculated before Isolation starts. The position of an occurrence of a term in an expression, can be represented as a list of numbers, which determine which parameter to take when traversing the expression from the dominant function (root) to the occurrence. For instance, the position of x in $\log_e x^2 -1$ is $<2, 1, 1>$, since it occurs in the 2nd parameter of log, the 1st of $-$ and the 1st of exponentiation (see figure 12-1). If expr-at(List,Exp) is the expression at position List in Exp then

expr-at($<2, 1, 1>$, $\log_e x^2 -1$) $\equiv x$

where $\equiv$ expresses identity between algebraic expressions.

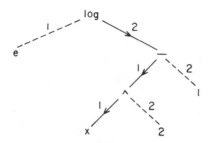

Figure 12-1: The Position of x

Exercise 37: What is the position of the 1 in $\log_e x^2 - 1$?

Exercise 38: Using the techniques of section 9.3 show that the selective application of Isolation rules will always terminate.

12.2.2 Collection

Similarly the Collection method can be programmed as the selective application of a set of rewrite rules. The Collection rule we used in our example was:

$$(U+V).(U-V) \Rightarrow U^2 - V^2$$

This has two occurrences of U on the left hand side and one on the right. Thus if U is matched to a term containing x then the number of occurrences of x will be reduced when the rule is applied. A similar argument holds for V. Thus this particular rule collects on both U and V. This is not true of the distributive law

$$U.V + U.W \Rightarrow U.(V+W)$$

which collects only on U.

The criterion for identifying a Collection rule is that it should contain some distinguished variable, U, which appears more often in the left hand side than the right. Thus if $occ(Var,Exp)$ is a function whose value is the number of occurrences of Var in Exp then a rule lhs $\Rightarrow$ rhs is useful for Collection relative to U iff $occ(U,lhs) > occ(U,rhs)$.

Another well known rule with this property is,

$$2.\sin U. \cos U \Rightarrow \sin 2.U.$$

This is not in its most generally useful form. For instance, as it stands it could not be applied to

3.sin x. cos x.

Thus it is better to use it in the form,

$$\sin U. \cos U \Rightarrow 1/2. \sin 2.U$$

where the smallest subterm on the left hand side, which contains all the occurrences of U, is isolated. We will call such a term, a *least dominating term* for U.

If we always arrange the Collection rules so that their left hand sides are least dominating terms in the variable they are useful for collecting then the application of the rules can be especially selective:

 – Collection need only try to rewrite terms which are least dominating in the unknown, e.g. it need only rewrite $(x+1).(x-1)$ in $\log_e (x+1).(x-1) = c$ (see figure 12-2);

 – the distinguished variable, U, must be matched to a term containing the unknown.

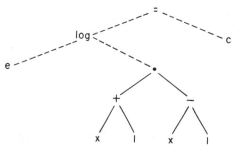

Figure 12-2: A Least Dominating Term in x

12.2.3 Attraction

The Attraction method is very similar to the Collection one. The sort of rewrite rules used by Attraction are:

$$\log_W U + \log_W V \Rightarrow \log_W U.V$$
$$W.U + W.V \Rightarrow W.(U+V)$$
$$(W^U)^V \Rightarrow W^{U.V}$$
$$U^{V.W} \Rightarrow (U^V)^W$$

In each of these the distinguished variables U and V are closer together on the right hand side than they are on the left.

What does it mean for two variables to be closer together in one place than in another? What is the notion of distance involved?

A simple measure of distance is easily defined by looking at the expression tree which contains the two occurrences. Consider figure 12-3. The shortest path between U and V in each of the trees is in solid type. In the tree of $\log_W U + \log_W V$ this path has length 4. In the tree of $\log_W U.V$ it has length 2. If we define the distance between two occurrences of terms as the length of the shortest path between them then U and V are closer in the second tree than in the first.

Thus a rule, lhs $=>$ rhs, is useful for attracting U and V if both U and V occur once each in lhs and rhs and if the distance between them in rhs is smaller than the distance between them in lhs. If we ever find a rule which attracts a whole bunch of variables at once then the definition of distance can easily be generalized to measure the smallest path connecting all the variables. As with Collection we will want lhs to be a least dominating term in U and V.

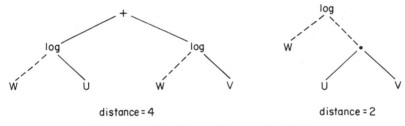

Figure 12-3: The Distance Between Two Terms

The selection criteria for applying these rules are also similar to those for Collection:

- the term to be rewritten should be a least dominating term in the unknown, x;

- the distinguished variables U and V should be matched to terms containing x;

- the undistinguished variables should be matched to terms not containing x (otherwise two occurrences of x may be moved closer together only to move two others further apart).

Notice that the same formula may be used as a rewrite rule by different methods, but because of the selection criteria it will be used in different ways. For instance,

$$W.U + W.V = W.(U+V)$$

is used by both Collection and Attraction. But whereas Collection will match W to a term containing x and reduce the number of occurrences of x, Attraction will match U and V to terms containing x and move the occurrences closer together. There is no reason why the same equation should not be used in different directions by different methods: the strong selection criteria will maintain termination.

In fact, the same formula can be used in different directions *by the same method* without loss of termination. Consider, the formula

$$U^{V.W} = (U^V)^W.$$

This is used in both directions by Attraction (see above rules), yet will not cause looping since, when used left to right W must be matched to a term free of x and when used right to left W must be matched to a term containing x.

12.3 Reasoning About Problems and Methods

Now we have an informal definition of the methods used in our example solution we can proceed to formalise the reasoning which was involved there.

12.3.1 Defining the Methods with Axioms

Suppose that the reasoning involved at each stage was something like:

Does the equation contain only a single occurrence of x? If so, Isolate it, if not then find a term which contains two occurrences of x and Collect these together. If this fails then try to Attract the two occurrences together. If any of these steps succeed then conduct the same reasoning on the resulting equation.

Let us pull out from this a particular piece of knowledge: that if an equation contains a single occurrence of the unknown then isolating this unknown will produce a solution to the equation. We have already seen how to express the information that an equation contains only a single occurrence of an unknown:

$$occ(x, \log_e x^2 - 1 = c) = 1$$

We will want to express the information that a formula is the result of applying the Isolation method to an equation to isolate an unknown. It will be convenient to do this in two bites, using the function, position, to say where the occurrence of the unknown is, e.g.

$$\text{expr-at}(<2,1,1>, \log_e x^2 - 1) \equiv x$$

and introduce the new function, *isolate*, which represents the result of isolating whatever is at a particular position in an equation.

$$\text{isolate}(<2,1,1>, \log_e x^2 - 1) \equiv (x = \sqrt{e^c + 1} \ \vee \ x = -\sqrt{e^c + 1})$$

Finally, we will want to be able to say that a formula is a solution, for an unknown, of an equation. Since there may be several solutions we will use a predicate, solve, and say, e.g.

$$\text{solve}(\log_e x^2-1=c, \ x, \ (x=\sqrt{e^c+1} \ \lor \ x=-\sqrt{e^c+1}))$$

Putting all this together gives us an axiom for reasoning about problems and methods, namely

1) $occ(X,A=B)=1$ & $\text{expr-at}(List,A) \equiv X$ &
$\text{isolate}(List, \ A=B) \equiv Ans \rightarrow \text{solve}(A=B,X,Ans)$

To get another axiom consider how the isolate function might be defined in terms of the applications of individual Isolation rewrite rules.

We will want to express the information that one expression is derived from another by applying an Isolation rewrite rule. There is only one such derived expression for each parameter position of the original expression, so we can introduce a function from parameter positions and expressions to expressions which we will call, isolate-rewrite, e.g.

$$\text{isolate-rewrite}(2, \ \log_e x^2 -1=c) \equiv x^2-1=e^c).$$

The Isolation method just consists of applying an Isolation rule to an equation and then entering the method recursively. This can be captured in the axiom:

2) $\text{isolate-rewrite}(N, \ Old) \equiv New$ & $\text{isolate}(Rest, \ New) \equiv Ans$
$\rightarrow \text{isolate}(cons(N,Rest), \ Old) \equiv Ans$

since the first parameter position to be isolated will be the number on the front of the position list of the unknown.

Exercise 39: This axiom only defines isolate when the expression Old is an equation. Give an axiom which defines isolate when Old is a disjunction of equations in terms of isolate on the disjuncts.

In a similar way we can design axioms for Collection and Attraction. The main differences being:

– since expressions can be collected and attracted in a variety of ways we will need to use predicates where we used functions above;

– neither Collection nor Attraction are applied repeatedly, like isolate, they are applied once and the resulting equation is solved.

For instance, an analogous axiom for Collection is:
3) $occ(X, \ Old) > 1$ & $\text{collect}(Old, \ X, \ New)$ & $\text{solve}(New, \ X, \ Ans)$
$\rightarrow \text{solve}(Old, \ X, \ Ans)$

where collect(Old, X, New) means that New is one result of collecting X in Old.

To complete the picture further axioms need to be written defining occ, position, isolate-rewrite, collect, etc, but we have enough to illustrate the principle of reasoning about problems and methods.

12.3.2 Searching for a Solution

If we add a goal clause defining a particular problem

$$\text{solve}(\log_e (x+1) + \log_e (x-1) = c, x, \text{Ans}) \rightarrow$$

and use this as top clause in a Lush Resolution derivation, then we can begin to see the shape of the search tree that would be developed. It is given in figure 12-4.

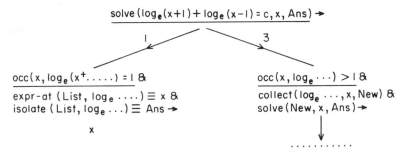

Figure 12-4: Search Tree for Reasoning about Problems and Methods

But how does all this help solve equations?

The answer is contained in the propositions involving isolate-rewrite (and similar predicates for Collection and Attraction) which state that such and such an expression is the outcome of applying a rewrite rule to an expression. The axioms which define these propositions define a rewriting process (see axiom 4 below). When the propositions are resolved away this rewriting process takes place as a side effect. Thus the reasoning at the level of problems and methods induces some reasoning at the level of manipulating algebraic expressions.

What is the advantage of organising things this way?

The branching rate of the tree in figure 12-4 is extremely small. Often there is no branching at all for long stretches of the tree. Useless branches get terminated quickly (as both the Collection and Isolation branches above are). The induced search at the level of algebraic manipulation, similarly involves very little search.

Thus an exhaustive search in which the axioms of algebra are applied indiscriminately to the equation to be solved is replaced by two well behaved searches running in close harness.

12.3.3 Meta Level Reasoning

What mathematical theory do the axioms 1, 2 and 3 above, belong to? It is not algebra, because the predicates solve, collect etc do not express relationships between numbers as all the algebraic predicates do, e.g. $=$, $>$ etc. solve and collect express relationships between algebraic formulae, i.e. they discuss the *representation* of algebra.

Whenever we reason about the representation of a mathematical theory we are said to be operating at the meta-level and the mathematical theory itself is said to be the object-level. Axioms 1, 2 and 3 above are axioms of the *Meta-Theory of Algebra*. The search tree in figure 12-4 represents a *meta-level search* and the algebraic search it induces is an object-level *search*. We will call this search control technique the technique of *meta-level inference*.

Notice that object-level predicates, like $=$, and object-level connectives, like v, are represented as meta-level functions, because they have to form terms which can be the parameters of meta-level predicates. We met this situation in the last chapter, when the object-level, LISP, equal predicate was represented as a meta-level function and the object level truth values, *tt* and *ff*, as meta-level constants. This discussion puts a theoretical gloss on the informal explanation in that chapter.

In fact there are two versions of $=$, $>$, v, etc: one in the object-level and one in the meta-level. We used the same symbol to represent both versions.

> *Exercise 40:* Go back to section 12.3.1. Mark each occurrence of an object-level $=$ sign with an o and each occurrence of a meta-level $=$ sign with an m.

Object-level variables must be represented as meta-level constants. Otherwise, unification can change the truth of propositions. Consider, for instance, the true proposition:

$$occ(x, \sin U) = 0$$

If unification should apply the substitution, $\{x/U\}$, then this will instantiate this true proposition into a false one, namely

$$occ(x, \sin x) = 0$$

However, if U is represented as a constant then no instantiation can take place.

As a consequence, the rewrite rules must all be represented as variable free formulae and matching and instantiating must all be defined as meta-level functions, e.g. the Isolation rules must be recorded as variable free assertions,

$\rightarrow$ isolate-rule(1, u$-$v $=$ w $=>$ u $=$ w$+$v)

and the definition of isolate-rewrite must involve match and instantiate,

4) isolate-rule(N, Lhs $=>$ Rhs) &
match(Lhs, Old) $=$ Theta &
instantiate(Rhs, Theta) $=$ New $\rightarrow$ isolate-rewrite(N, Old, New).

12.4 Summary

Bob Welham has embodied the technique of meta-level inference in a program, PRESS, for solving symbolic, transcendental equations [Bundy and Welham 81]. PRESS is currently being developed by a team of researchers, led by the author. This chapter is a rational reconstruction of the PRESS program. We can summarize the technique as follows.

- The problem to be solved is analysed using meta-level concepts, like occ and position.

- This analysis is used to access an appropriate method of solution, e.g. Isolation or Collection.

- The entire process can be thought of as inference at the meta-level inducing inference at the object-level.

In the case of PRESS the object-level inference consists of the application of rewrite rules to an expression.

This technique generalizes to other domains. The final formalization of the group theory example of chapter 7 was a meta-theoretic axiomatization. We designed axioms like:

$\rightarrow$theorem((i))

$\rightarrow$common-technique((i), abelianness)

$\rightarrow$proof((i), aob $=$ eoaob $=$.....$=$ boa)

These axioms could have played a part in inducing an object-level inference process, where the object-level was group theory.

In fact, the reasoning of human mathematicians is usually at the meta-level of the mathematical theory they are working on, so that this technique is often a good way to model their thought process.

When I wrote, in chapter 7, that the standard axiomatizations of mathematical theories, provided by Mathematical Logic, were not the best for automating mathematical reasoning, it was meta-theoretic axiomatizations that I had in mind as a non-standard alternative. I hope that this alternative will receive considerable attention in the future.

Further Reading Suggestions

[Bundy and Welham 81] is the best general introduction to PRESS. [Borning and Bundy 81, Bundy and Silver 81, Bundy and Sterling 81, Sterling et al. 82] describe recent extensions.

Part IV:
Mathematical Invention

13
Concept Formation

☐ This chapter describes how a computer can form definitions and conjectures.

☐ Section 13.1 gives examples of concepts being formed.

☐ Section 13.2 analyses these examples to identify various concept formation processes.

☐ Section 13.3 formalizes these processes and the knowledge they operate on. Concept formation is described as inference.

☐ Section 13.4 describes how to control this inference using heuristic search.

☐ Section 13.5 describes the performance of a program, AM, based on these ideas.

Up to now this book has concerned itself with only one aspect of mathematical reasoning: the proving of theorems. But this is not the only kind of mathematical reasoning indulged in by human mathematicians. In the next two chapters we consider two of these other aspects, namely

- the making of interesting definitions and conjectures
- and the formation of a mathematical model from an informal problem specification.

This chapter is a rational reconstruction of a program, AM, written by Doug Lenat [Lenat 82], for making interesting definitions and conjectures, a process we will call *concept formation*. That reasoning is involved in these activities is illustrated by the following two examples of the sort of thing that AM can do. The first example shows how AM discovered for itself the concept of prime number and the second example shows how AM went on to conjecture the Prime Unique Factorization Theorem.

13.1 How Definitions and Conjectures Can Be Made

- Given times, a function from bags* of numbers to numbers:

- Find some examples of times, e.g. times([2,1,2]) = 4;

- Create inv-times, a function from numbers to sets of bags of numbers, by forming the inverse of times; (AM is using the trick discussed in section 4.3.1 for ensuring that inv-times is a function by forming a set of values.)

- Find some examples of inv-times, e.g.

*See section 4.3.3 for a definition of bags.

inv-times(4) = {[2,1,2], [1,4], [4], [2,2]}; (In order that this set should not be infinite, AM only allowed one occurrence of the unit element, 1, in each bag.)

- Create divisors, a function from numbers to sets of numbers, by composing inv-times and generalized union, i.e. by applying inv-times and then taking the union of all the bags in the resulting set;

- Find some examples of divisors, e.g. divisors(4) = {1,2,4};

- Create primes, the set of those numbers whose divisors are doubletons, e.g. divisors(3) = {1,3}.

Notice how the reasoning in this example consisted of a succession of creating new concepts from old and finding examples of the new concepts. As we will see, the finding of examples, while not strictly necessary to the progression of definitions, does help AM to decide what definitions are interesting to make.

In order to make interesting conjectures, a third sort of operation is required: the checking of the examples in an attempt to discover regularities and patterns. This is illustrated by the next example.

- Given inv-times, a function from numbers to sets of bags of numbers:

- Find some examples of inv-times, e.g.
  inv-times(4) = {[2,1,2], [1,4], [4], [2,2]};

- Check examples of inv-times and notice that inv-times always contains a bag of primes, e.g. [2,2] ∈ inv-times(4);

- Create prime-times, a function from numbers to sets of bags of primes, by restricting the range of inv-times to bags of primes;

- Find some examples of prime-times, e.g.
  prime-times(4) = {[2,2]};

- Check examples of prime-times and notice that all examples of prime-times are singletons;

- Conjecture the Unique Factorization Theorem, namely that prime-times will always be a singleton.

13.2 Operations of Concept Formation

The operations that were involved in the above two examples of concept formation were: creating new concepts (e.g. functions, sets, etc), finding examples of concepts, checking examples of concepts and making conjectures. We will consider each of these in more detail.

13.2.1 Creating New Concepts

New concepts were created from old ones above, in a variety of ways.

- inv-times was created from times by a process of *inversion*, that is
  $$\text{inv-times}(X) = \{Y : \text{times}(Y)=X\}.$$
  As noted in section 4.3.1, this particular way of forming inverses guarantees that they are functions: inv-times(X)=Y $\longleftrightarrow$ times(Y)=X would not be a proper definition unless there was only one such Y for each X.

- divisors was created from inv-times by *composition* with generalized union (gen-union), that is
  $$\text{divisors}(X) = \text{gen-union}(\text{inv-times}(X)).$$

- prime-times was *created* from inv-times by restriction, that is
  $$\text{prime-times}(X) = \{Y : Y \in \text{inv-times}(X) \ \& \ \text{bag-of-primes}(Y)\}$$

- In addition, AM can also form new concepts by *coalescing* old ones, i.e. by identifying their parameters. For instance, it forms the function, square, from binary multiplication by identifying its parameters, i.e.
  $$\text{square}(X) = X.X$$

- And AM can form new concepts by *structural modification* of definitions of old ones, e.g. it drops a conjunct from the definition of equal to form a new definition – of the concept equal-length.

13.2.2 Finding Examples of Concepts

Having defined a concept AM tries to find examples of it. This enables AM to decide how important the concept is and to suggest new concepts and conjectures that might be made. It uses a variety of ways of finding examples. Some of these are listed below.

- *Instantiating the definition.* For instance, given that AM has an example of times, it might insert it in the definition of inv-times to generate an example of that. This is how the examples above were generated.

- *Look at the concept's generalizations.* For instance, given that AM has several examples of numbers, and that it knows that numbers are a generalization of primes, then each of the example numbers is a candidate for an example prime.

- *Run calculation procedures.* If AM has a function whose range is the concept in question and if there is a calculation procedure associated with that function and if examples of the domain of the function are known, then the procedure can be run to generate examples of the concept. For instance, suppose AM were trying to find examples of sets. It knows that divisors is a function from numbers to sets of numbers and it knows some numbers. If it also has a procedure for calculating divisors then this can be run on the numbers and will produce some example sets.

13.2.3 Checking Examples of Concepts

AM checks these examples periodically for a variety of reasons. We have met one of these, namely to look for regularities and patterns among them.

- This may suggest the creation of a new concept, as the noticing that inv-times always contained a bag of primes led to the definition of prime-times,

- or it may suggest the making of a conjecture, as the noticing that prime-times was always a singleton led to the conjecture of the Unique Factorization Theorem.

Another role of checking is to vet the examples. Some of the methods of obtaining examples, like the use of examples of generalizations, are liable to error. Thus some 'examples' may need to be rejected or modified.

Lastly, if there are just not enough examples, the checking operation may suggest finding some more. How it does this we discuss below.

13.2.4 Making Conjectures

It is one thing to say 'look for regularities' among the examples, but what does this mean in practice: what sort of regularities should be checked? AM looks for four kinds of regularities and these are listed below.

- One concept is an example of another, e.g. AM notices that all the examples of prime-times are singletons and hence conjectures that this is always the case.

- One concept is (almost) a specialization of another, e.g. AM notices that all the examples of primes were also examples of odd-primes, except the boundary prime 2. It conjectures that all primes except 2 are odd.

- One concept is (almost) the same as another, e.g. AM notices that an operation it had just defined was the same as times. It merges the two concepts and boosts its interest in the resulting concept.

- One concept is related to a second by a third, e.g. AM notices that bag-of-primes is related to inv-times by set membership, that is inv-times always contains a bag of primes. It restricted inv-times to the range, bag of primes, to define the new function prime-times. Since new relations are defined from time to time, the notion of 'is related' is continually being broadened.

13.3 Formalizing the Knowledge

In order to get a computer to perform these operations on concepts it is necessary to formalise the knowledge involved. In the discussion above we have seen the need for all sorts of knowledge about the concepts of mathematics.

Firstly, since AM's purpose is to make definitions and conjectures it needs to be able to record these. This can be done using predicates, 'definition' and 'conjecture', between a particular definition or conjecture and the name of the concept, e.g.

definition(primes, {X: divisors(X) ∈ doubletons})

conjecture(primes, prime-times(X) ∈ singletons)

You might think that 'definition' should be a function, but AM sometimes records several alternative definitions for a concept, so a predicate is appropriate.

As we have seen, AM also finds examples of concepts, so it needs to be able to record these. We can do this using a predicate, 'example', between the example and the name of the concept, e.g.

example(numbers, 2)

example(sets, numbers)

example(times, times([2,2]) = 4)

To help AM find examples of concepts, some additional information must be recorded. For instance, if generalizations of a concept are to suggest candidate examples, then the generalizations of the concept must be stored. If examples are to found by running calculation procedures of functions whose range is the concept, then that function's calculation procedure, domain and range must all be recorded, e.g.

generalization(primes, numbers)

generalization(odd-primes, primes)

procedure(times, 'a computer program')

type(times, bags-of-numbers ↦/ numbers)

For reasons of efficiency and presentation, AM recorded all the above information in the form of a table (or frame see appendix III. It also stored the information redundantly, e.g. the table for numbers records *both* its generalizations and its specializations, the table for times and numbers both record the fact that numbers are the range of times. An example table is given in Table 13-1.

Table 13-1: Record of Information about Concepts

name	numbers	times
definition	{X: bag-of-ts(X)}	parallel-join2(bag,bag,proj2)
conjecture	intersect(X,Y)	associative
	ε numbers	
example	0, 1, 2, 3	times([2,2])=4
isa	object	operation
generalization	sets	parallel-join2
specialization	primes	square
procedure	—	'a computer program'
type	numbers	bags-of-numbers ↦ numbers
in-domain-of	inv-times	—
in-range-of	times	—

Note that specialization is the inverse of generalization, isa is the inverse of example and in-domain-of and in-range-of are both inverses of type. The relation between specialization and isa is analogous to that between set inclusion and set membership, e.g. all primes are numbers and both primes and numbers are examples of objects. AM employs a representation language to make this redundant storage efficient. Knowledge is entered by typing in data structures, such as 13-1, and the representation language automatically adds or changes the redundant inverse slots.

13.3.1 Initial Concepts

In the same way that an automatic theorem prover needs a set of axioms as a basis before it can derive any theorems; AM needs a set of initial concepts before it can derive new ones and make conjectures. AM's initial concepts were chosen without any deep analysis of what would be parsimonious or psychologically plausible, but they reflect the sort of knowledge that a preschool child might be expected to have. They are represented diagrammatically in figure 13-1.

In this diagram specialization relationships are represented by single links and example relationships by treble links.

AM has a represention, not only of sets, lists and bags, but also of *osets*, which are ordered like lists but, like sets, have no multiple elements. These four kinds of structure form a little quartet.

The operations of inversion, restriction, composition, coalescing and structural modification are all represented explicitly as concepts. Thus, AM can form the inversion of times (inv-times) by composing inversion and times; or the restriction of inv-times to primes (prime-times) by composing restriction and inv-times.

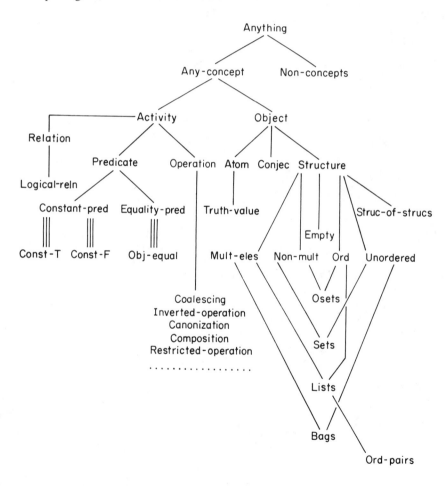

Figure 13-1: AM's Initial Concepts

13.3.2 Formalizing Operations

Just as the knowledge about the initial concepts can be represented as axioms, the operations of finding examples and making definitions and conjectures can be represented as rules of inference.

For instance, the fact that all known examples of one concept are examples of another can be represented by the formula,

$$\forall Ex \; example(C1, Ex) \to example(C2, Ex),$$

where C1 is the first concept and C2 the second. In this case AM can conjecture that C1 is a specialization of C2, i.e.

$$conjecture(C1, C1 \subseteq C2).$$

We can represent this operation as a rule of inference, i.e.

$$\frac{\forall Ex \; example(C1, Ex) \to example(C2, Ex)}{conjecture(C1, C1 \subseteq C2).} \qquad (i)$$

AM makes new definitions by inverting, restricting, composing, coalescing, structurally modifying, etc old definitions. Since all these operations are represented explicitly as initial concepts AM need only use composition. Thus the making of new definitions can be represented by the rule,

$$\frac{definition(C1, D1(X)) \quad definition(C2, D2(Y))}{\exists C3 \; definition(C3, D2(D1(X)))}$$

If AM knows of a function, F, from $D \mapsto R$ and it knows an example of concept D and it has a calculation procedure associated with F then it can find an example of R by running the procedure on the example of D. Let run(Proc, Ex) represent the result of running program, Proc, on input, Ex, then this operation can be represented by the rule of inference,

$$\frac{type(F, D \mapsto R) \quad procedure(F, Proc) \quad example(D, Ex)}{example(R, run(Proc, Ex))}$$

Exercise 41: AM knows that any example of a concept is an example of a generalization of the concept. Represent this operation as a rule of inference.

13.4 Concept Formation as Heuristic Search

Thus concept formation can be viewed as a forward search process, by applying the above rules of inference to AM's initial concepts. Like many of the search processes we have considered in this book, there are many possible applications which can be made at every stage: if the search process is not to become bogged down in a combinatorial explosion there must be some guidance.

Lenat's solution to this problem was to use meta-level inference* (cf chapter 12). AM reasons about the representation of concepts that it has derived so far: 'Which slots in a concept's table have been filled in?', 'Which slots are empty?', etc. On the basis of this reasoning AM suggests new tasks: 'Fill in this slot on this concept.', 'Check the contents of this slot.'. As these tasks are achieved new tasks are suggested and the representation of mathematical concepts is gradually expanded.

To conduct this meta-level inference AM uses a systems of meta-level rules.** Examples of such rules are:

29. To fill in examples of X, where X is a kind of Y, Inspect the examples of Y; some of them may be examples of X as well. The further removed Y is from X, the less cost-effective this rule is.

48. If the totality of examples of concept C
is too large to be interesting,
then consider these three possible reactions:

(i) specialize C;
(ii) forget C completely;
(iii) replace C by one conjecture.

We follow Lenat by writing these meta-rules in English. The meta-rules fall roughly into two classes: those, like the first example above, which enable a particular task to be achieved; and those, like the second example, which suggest new tasks. The meta-level inference process can also be visualised as a forwards search of a tree, in which the nodes correspond to the 'check' or 'fill in' tasks and the arcs are labelled by meta-rules.

In order to guide the growth of this task tree (see figure 13-2), AM used the technique of heuristic search. Each task was assigned a numeric score. As each new task was suggested its numeric score was calculated and this determined its position on an agenda. When AM had finished its current

---

*Although he did not call it this and, indeed, prefers not to distinguish meta-rules from rules.
**Called 'heuristic rules' or 'heuristics' by Lenat.

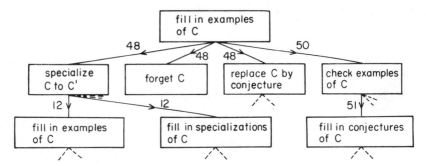

Figure 13-2: AM's Development Viewed as Growing a Tree of Tasks

task and was ready to start a new one, it would take off the front of the
agenda, the task with the best score.

To calculate the numeric score of a task, AM combined numeric values
from various aspects of the task, namely

- A value, R, is inherited from the meta-rule which originally put the
  task on the agenda, or, R_i , from each distinct meta-rule, if several
  meta-rules suggested the same task.

- A symbolic 'reason' is also inherited from the meta-rule. AM uses
  these 'reasons' to ensure that the a meta-rule which is fired several
  times only contributes its R value to a task once. (The 'reasons' are
  also printed to the user to explain AM's current activity.)

- A value, C, is associated with the concept whose table is being
  worked on and

- a value, S, with the sort of slot which is being filled in or checked.

- A different value, A, is given depending upon whether the current
  task is to 'fill in' or to 'check'.

These values are combined to give an overall score for the task using the
formula:

$$|| \sqrt{\textstyle\sum_i R_i^2} ||*[0.2*A + 0.3*S + 0.5*C]$$

Lenat found that the performance of AM was unaffected by slight variations
to this arithmetic formula, suggesting that little hangs on its precision.

The R values of this formula are stored with each meta-rule. The values A
and S are global parameters associated with 'fill in', 'check' and the slot
names. The C value must be associated with each concept and this is most
conveniently done by storing it in that concept's table as the value of a slot

called the concept's *worth*.

When a new concept is created it must be assigned a worth value for use in future calculations of a task's score. The worth of a new concept is calculated from the worths of the concepts which define it. Thus, when inv-times is created, its worth is calculated from the worths of inversion and times. The formula used to make this calculation is local to the particular concepts involved. Information about how to do the calculation is stored on yet another slot of each concept's table, the *interest* slot. Some of AM's meta-rules can also change the worth of a concept and also the score of a task.

13.5 The Performance of AM

AM started with the 76 concepts in figure 13-1. By running its 242 meta-rules to suggest and achieve about 256 'fill in' and 'check' tasks, it gradually filled out the representation of these concepts and defined and filled out new concepts. After about an hour of cpu time AM had about 300 concepts. Of the new concepts defined about 25 were really interesting and about 100 more were acceptable. The remainder were worthless. The amount of information stored about the original concepts approximately tripled.

Among the interesting new concepts developed by AM were: numbers, prime numbers and maximally divisible numbers. This latter kind of number is the dual of the prime numbers: each maximally divisible number having more divisors than any smaller number. Lenat originally dismissed maximally divisible numbers as uninteresting, but later took them more seriously and managed to prove some theorems about them. Finally, Lenat discovered that his theorems had previously been discovered by the famous Indian mathematician, Ramanujan.

Among the interesting conjectures made by AM were: de Morgan's Law, the Prime Unique Factorization Theorem and Goldbach's Conjecture. The definition of primes and the proposing of the Unique Factorization Theorem were given as examples in section 13.1.

In [Lenat 82] the trace of AM which is claimed to be the proposing of Goldbach's Conjecture is in fact a weaker conjecture, namely that every even number is the sum of some number of primes, rather than just two primes. This weaker conjecture is trivial to prove, e.g. $6 = 2+2+2$, etc. AM did propose Goldbach's Conjecture on a previous run and Lenat was misled by the similarity of the printout.

Lenat performed various experiments with AM: changing the way that the tasks were selected and varying the initial concepts. AM was remarkably robust under changes of the method of task selection. Slight changes produced almost no difference in AM's behaviour; more radical changes slowed down the development of interesting new concepts by a factor of three, but did not prevent that development.

AM's meta-rules were shown to be robust by exchanging the set-theoretic initial concepts for geometric ones, i.e. the concepts of point, line, angle, triangle, etc. AM was still able to make reasonably interesting new definitions and conjectures, e.g. the congruence and similarity of triangles. Although some of the geometric results had a strong arithmetic flavour.

In all the experiments, AM's concept formation slowed down after about an hour of cpu time. The problems were twofold.

- – As the number of concepts grew and the amount of information stored about each one increased, the computer's memory was gradually exhausted.

- – The quality of the new information being discovered began to degrade.

Lenat attributes this degradation of quality to the fact that to discover interesting information about a concept, meta-rules are required especially geared to that concept. The new concepts discovered by AM are more and more specialized, but it cannot also discover new meta-rules geared to these special concepts. It must, instead, rely on very general purpose rules and this becomes an increasing liability.

13.6 Summary

It is possible to represent the formation of new mathematical concepts and the making of conjectures as a process of inference. This inference process is extremely explosive, but can be guided by meta-level inference, which in turn can be guided by heuristic search. The proof that this is a practical proposition is embodied in a computer program, AM.

Starting from a base of very general concepts, AM was able to discover many interesting arithmetical concepts and make several interesting conjectures. AM is guided by a simple numeric model of 'interestingness', in which the 'worth' of a concept is inherited from the concepts from which it was defined. AM did not use any notion of proof to confirm its conjectures or to determine interestingness, but this would be an interesting extension of the AM ideas.

Further Reading Suggestions

[Lenat 77a] is a short introduction to AM. [Lenat 82] is longer and more detailed account. [Lenat 77b] is a more general discussion of the issues.

14

Forming Mathematical Models

☐ This chapter describes how mathematical problems, stated in English, can be solved by translating them into equations.

☐ Section 14.1 describes and criticises the simple technique of keyword replacement used in the STUDENT program.

☐ Section 14.2 describes the intermediate representation used, in the MECHO program, as a staging post between the English and algebraic representations of a problem.

☐ Section 14.3 describes the role of inference in bridging the gap between the information in the intermediate representation and that required to form the equations.

☐ Section 14.4 describes the formalization of physical laws.

☐ Section 14.5 describes the process, used by MECHO, to form equations from the intermediate representation.

☐ Section 14.6 summarises the use of meta-level inference in MECHO.

In this chapter we consider another 'non-theorem proving' aspect of mathematical reasoning: the formation of mathematical models from an informal specification. In particular, we will consider how the description of a problem in English can be translated into a system of mathematical formulae, e.g. equations.

14.1 Keyword Replacement

One class of problems, sometimes called *Algebra Word Problems*, (Not to be confused with the word problems of group theory, ring theory, etc defined in section 9.6.3.) admit of a particularly simple solution. Various key words in the English statement of the problems can be translated directly into mathematical symbols. Phrases which describe objects, called *noun phrases*, can then be translated into variables, so that each clause gives rise to an equation.

The translation rules are as follows:

– the keyword 'is' is replaced by '=';

- noun phrases in which a number is followed by a unit, e.g. '5 inches', are replaced by the multiplication of the number and the unit, e.g. '5.inches';
- some special phrases, e.g. 'X percent less than', have special translations, e.g. '$(100-X)/100$.';
- questions and commands, e.g. find X, what is X, have the special effect of marking X as a variable to be solved for;
- any remaining noun phrases are replaced by variable names: identical or similar noun phrases being replaced with the same variable.

Consider, for example, the problem,

The price of a radio is 69·70 pounds. If this price is 15 percent less than the marked price, find the marked price.

We will translate this problem statement clause by clause.

The first clause is,

The price of a radio is 69·70 pounds.

'is' translates to '=' and '69·70 pounds' to '69·70.pounds'. This leaves the noun phrase 'The price of a radio' which we will replace by a new variable, price. Thus the whole clause translates to,

price = 69·70.pounds

The second clause is,

this price is 15 percent less than the marked price

Again 'is' translates to '=' and the special phrase '15 percent less than' to '·85.'. This leaves the noun phrases 'this price' and 'the marked price'. A procedure is needed which can recognise that the first of these refers to 'the price of a radio' while the second is a new variable, 'marked-price'. The problem this procedure has to solve is called the *noun phrase reference problem*. We will be meeting it again. Assuming that this procedure is provided, the translation of the second clause is,

price = ·85.marked-price

The final clause is,
Find the marked price.

This is a command, and so has the effect of marking 'marked-price', as a variable to be solved for.

The equations can now be solved for 'marked-price'. Since they are linear simultaneous equations a procedure to solve them can be easily provided. This will yield the answer,

marked-price $= 82.$ pounds.

To finish off the process this equation should be translated back into English by applying the rules above in reverse. Thus '=' should be replaced by 'is'; '82.pounds' by '82 pounds' and 'marked-price' by 'The marked price'. This will give the clause,

The marked price is 82 pounds.

> *Exercise 42:* Translate into an equation the clause: 'The distance between London and Edinburgh is 400 miles'.

A program based on these techniques, called STUDENT, has been built by Danny Bobrow [Bobrow 64]. It can solve a wide range of simple algebra word problems, like that above. It does, however, have its limitations, for instance,

- it can only deal with a limited subset of English: verbs like 'is', 'find', etc connecting a small range of noun phrases;

- its procedure for solving the noun phrase reference problem is successful only in simple cases, where the similar phrases share some common string of words;

- it can only solve sets of linear simultaneous equations.

These problems could have been solved by a process of gradual improvements and extensions, however, STUDENT had a more fundamental limitation, and this is illustrated by the following example.

Two particles of mass m1 and m2 lbs are connected by a light string passing over a smooth pulley. Find the acceleration of the particle of weight m2 and the tension of the string supporting it.

Figure 14-1 is a diagram of the situation being described here.

Figure 14-1: A Pulley System

There are noun phrases in this problem specification which cannot be translated into equations, or any other kind of algebraic formula, e.g.

> Two particles .... connected by a light string passing over a smooth pulley.

But this is not an optional part of the specification which can be ignored: it sets the scene for the algebraic information. STUDENT has no way of representing such information.

Furthermore, the problem cannot be solved without a piece of 'scene setting' information, which is not provided in the problem statement, but is expected to be assumed by the problem solver, namely that the string is hanging vertically on either side of the pulley. The problem solver is expected to deduce this from his knowledge of gravity and freely hanging objects or to assume it from his familiarity with pulley systems in Mechanics problems.

Thus in order to solve problems like the pulley one above an automatic problem solver must be able to represent 'scene setting' information provided in the English statement and common-sense information about gravity, etc. A representation of the problem is needed which is intermediate between the English statement and the sets of equations. The English statement must be translated first into this intermediate represent-ation, where the common-sense knowledge can act on it and fill out the problem statement; and then this intermediate representation must be translated into sets of equations. This strategy is diagrammed in figure 14-2.

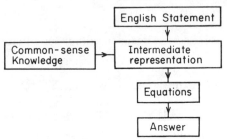

Figure 14-2: Strategy for Problem Solving

A Mechanics problem solving program, called MECHO, has been built by the author and others [Bundy et al 79a]. It can solve the pulley problem above and a variety of problems from other areas of Mechanics. This chapter is a rational reconstruction of the MECHO program.

We will deal mainly with the formalization of the intermediate repre-sentation and of the common-sense knowledge, and with the procedure for extracting equations from the intermediate representation. The problem of translating from English into an intermediate representation is called *Natural Language Understanding.* The study of this problem is a separate area of Artificial Intelligence. It has come a long way from the simple keyword replacement techniques described earlier and requires a book all of its own. The interested reader is refered to [Winograd 72, Mellish 80].

14.2 Formalizing the Intermediate Representation

The scene described in the pulley problem above consists of a configuration of four objects. The intermediate representation must represent this configuration and associate various physical quantities with them. There is an element of arbitrariness about any representation of such configurations, that is there are many choices of predicates, functions and constants to be made and most of the alternatives are equally acceptable. Where a particular choice is important we will point this out below.

The four objects – the two particles, the string and the pulley – can be represented by the constants, part1, part2, str and pull, respectively. We will want to record what sort of object each one of these constants represents, and we can do this with a predicate, *isa*, e.g.

isa(part1, particle)
isa(part2, particle)
isa(str, string) and
isa(pull, pulley)

Every object has exactly one type, so we could have made isa a function. In fact none of the relationships in the intermediate representation will be represented by functions. There are several reasons for this design choice, one of which was first described in section 4.3.2: that using no functions causes the arguments to be essentially boolean. We will meet further reasons in section 14.6, where we will also see how the missing functional information can be reintroduced into the problem solving process.

Now we must represent the relationships between these objects. These can be represented at various levels. At the highest level the four objects form a very common configuration in Mechanics problems – a pulley system. Thus the relationship might be represented by a single predicate, *pulley-sys*, e.g.

pulley-sys(pull, str, part1, part2).

At a lower level the configuration can be represented as a series of contact relations between the point objects and parts of the string. In problems of this kind the particles and the pulley can be idealised as 0 dimensional objects and the string as a 1 dimensional object. We need to specify three points on the string: the two ends and the point of contact with the pulley. This can be done with the predicates *end* and *contact-pt*, i.e.

    end(str, end1, left)
    end(str, end2, right)
    contact-pt(str, cpt)

Contact relations can then be asserted using the predicate *contact*, i.e.

    contact(end1, part1)
    contact(end2, part2)
    contact(cpt, pull)

To complete this low level picture the shape of the string needs to be described. It is divided into two parts. The part supporting part1 we will call bit1 and the part supporting part2 we will call bit2. The relationship between these bits and the whole string can be captured with the predicate, *partition*,

    partition(str, <bit1, bit2>)

where the second parameter is a disjoint list of subparts of the first, and the order of this list indicates the left-right order of the subparts.

Finally, bit1 and bit2 must be asserted to be straight with vertical inclination. This can be done with the predicates *incline* and *concavity*. Both of these were developed to express the shape of curved 1 dimensional objects: hence *incline* gives the angle between the tangent at a point and the x-axis; and *concavity* would normally say on which side a curve was concave, but here takes the special value of stline.

    incline(bit1, end1, 90)
    incline(bit2, cpt, 270)
    concavity(bit1, stline)
    concavity(bit2, stline)

Here the inclination of bit1 is directed from end1 to cpt and the inclination of bit2 is directed from cpt to end2 (see figure 14-3). This asymmetry is introduced to simplify the task of representing the asymmetry of the direction of travel of the two particles, i.e. if one goes up then the other goes down.

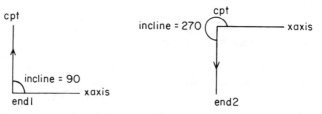

Figure 14-3: The Inclination of the String Parts

In practice both high and low level relationships are required. Some of the lower level relations can be recovered by translating the English statement of the problem, e.g. the *contact* and *isa* relations. These relationships can be used as cues for the higher level, *pulley-sys*, relationship and the remaining low level relationships, e.g. the string shape relations can be inferred from this. The high level, *pulley-sys*, proposition is an example of a *schema*. This word was chosen following Bartlett [Bartlett 67] who first used it to describe the behaviour he observed when people recognised an object: a few properties of the object are perceived and these are used to cue an object schema, whose remaining properties are then verified by further perception.

In the case of Mechanics problems the schemata correspond to standard configurations (pulley systems) or situations (motion in a straight line). Schema cueing allows the unstated assumptions of the problem (called 'house rules' in [Marples 74]) to be brought in. In addition to the shape of the string these will also include assumptions about physical quantities, e.g. that the string is inextensible unless otherwise stated. The schema cueing mechanism has to allow all these default assumptions to be overruled by contradictory information in the problem statement. For instance, if a non-zero coefficient of extensibility for the string were given then the default assumption that the string is inextensible would be overruled.

This completes the description of the configuration, so we go on to consider how physical quantities, masses, accelerations, etc, can be associated with the various objects. For instance, a mass of m1 lbs must be associated with part1 and m2 lbs with part2. A particle has only one mass, but this may be expressed in various units, lbs, stones, grams etc. Translation between units may be necessary during the solution of the equations, so the mass of a particle is expressed in two bites: one asserting a unique mass and the other expressing this in terms of a number of units, i.e.

mass(part1, mass1)
measure(m1, lbs, mass1)
mass(part2, mass2)
measure(m2, lbs, mass2)   (i)

A similar technique can be used to represent the tension of the string.

tension(bit2, tsn)
measure(t, lbs.ft/sec^2, tsn)   (ii)

and the acceleration of the particles

accel(part2, acc, 90)
measure(a, ft/sec^2, acc)   (iii)

but since the coefficients of friction and extensibility are dimensionless quantities no measure proposition is required.

friction(pull, 0)
extensibility(str, 0)

Exercise 43: Represent the information that the mass of a stone is 5 oz
and its acceleration is 32 ft/sec^2 downwards.

Finally, the status of the various physical quantities has to be represented
so that MECHO knows what it is supposed to solve for in terms of what.
This is done using two meta predicates, *sought* and *given*, which specify
which quantities are to be solved for and which may appear in the answer,
respectively. Thus the pulley problem above yields the propositions,

sought(acc)
sought(tsn)
given(mass1)
given(mass2)

14.3 Bridging the Gaps

The propositions above constitute the intermediate representation of the
problem. This is extracted from the English problem statement by natural
language understanding programs. But in order to solve the problem we
have seen that additional information is required: we need to bridge the gap
between the information provided in the problem statement and that
needed to solve the problem.

We have already seen one 'bridging' technique – the cueing of schemata.
But this technique does not solve all the problems. For instance, to solve the
problem it is necessary to resolve forces about both particles. To do this
MECHO needs to know the acceleration and mass of both particles and the
tensions in the bits of string which they are connected to. This information is
recorded above for part2 (equations (iii), (i) and (ii)), but vital parts of it are
missing for part1, because they were not mentioned in the English statement
of the problem.

However, the missing information about part1 can easily be recovered by
inferring it from the information about part2. All that is required are
inference rules of the form:

tension(Bit2, Tsn) & partition(Str, <Bit1, Bit2>) &
extensibility(Str, 0) & pulley-sys(Pull, Str, P1, P2) &
friction(Pull, 0)
$$\rightarrow \text{tension(Bit1, Tsn)}$$

which says that the tension in one bit of the string of a pulley system is the same as the tension in the other bit, provided that the string is inextensible and the pulley is frictionless.

accel(Part2, Acc, Dir2) &
pulley-sys(Pull, Str, Part1, Part2) & extensibility(Str, 0) &
end(Str, End1, left) & incline(Str, End1, Dir1)
$\quad\quad\quad$→ accel(Part1, Acc, Dir1)

which says that the acceleration of one particle in a pulley system has the same magnitude as the other, provided that the string is inextensible.

The schema cueing mechanism can also be implemented as an inference proccss in two stages, namely:

– the cueing of the schema;

– and the use of this to infer default information.

The cueing stage consists of the application of axioms like:

isa(Part1, particle) & isa(Part2, particle) &
isa(Pull, pulley) & isa(Str, string) &
end(Str, End1, left) & end(Str, End2, right) &
contact-pt(Str, Cpt) &
contact(End1, Part1) & contact(End2, Part2) &
contact(Cpt, Pull)
$\quad\quad\quad$→ pulley-sys(Pull, Str, Part1, Part2)

Thus when sufficient evidence has accumulated about the existence of a pulley system the high level proposition asserting the existence of one can be deduced. There may be alternative sets of sufficient evidence which would confirm the existence of a pulley system. This is reflected by having additional axioms which differ from the one above by having different conditions on the left hand side of the implication arrow.

When axioms of the above sort have been used to assert the existence of a schema it can be used to infer default information. This is done with axioms like:

pulley-sys(Pull, Str, Part1, Part2) → extensibility(Str, 0)
and
pulley-sys(Pull, Str, Part1, Part2) →
$\quad\quad$∃Bit1 ∃Bit2 partition(Str, <Bit1,Bit>)

The fact that this is *default* information means that these axioms must be treated in a special way by the inference system: they must not be used if contradictory information is already known. The MECHO inference system has such axioms specially marked. Before it uses them it tries to see if a contradictory conclusion is already known. In the first example above it sets up the subgoal

extensibility(Str, X)

if this succeeds then the default axiom is never used.

> *Exercise 44:* Draw the Lush Resolution search tree for the following clauses, using (vii) as top clause.
>
> (i) accel(Part2, Acc, Dir2) &
>   pulley-sys(Pull, Str, Part1, Part2) &
>   extensibility(Str, 0) &
>   end(Str, End1, left) & incline(Str, End1, Dir1)
>   → accel(Part1, Acc, Dir1)
>
> (ii) pulley-sys(Pull, Str, Part1, Part2)
>   → extensibility(Str, 0)
>
> (iii) → pulley-sys(peg, rope, man, bucket)
>
> (iv) → accel(bucket, quickly, up)
>
> (v) → end(rope, knot, left)
>
> (vi) → incline(rope, knot, down)
>
> (vii) accel(man, quickly, down) →

14.4 Extracting Equations from the Intermediate Representation

We have seen how all the information in the problem statement can be represented as propositions and how gaps in this information can be bridged by inference rules and schema cueing, but how can a mathematical model in the form of equations be formed from this intermediate represention?

Physical formulae, like the 'resolution of forces' formula, F=MA, must clearly be brought to bear. But what is the relation between the F, M and A of the formula and the propositions of the intermediate representation? M is the mass of whatever object we are resolving about, e.g. if MECHO decides to resolve about Object then

mass(Object, M)

The situation for A is more complicated: A depends both on the acceleration of Object and on the direction, Dir, in which MECHO decides to resolve, i.e.

$A = \text{Acc.cos(Dir}-\text{Dir1)}$
      where accel(Object, Acc, Dir1)

And the situation for F is most complicated of all since F is the sum of all forces acting on Object.

MECHO represents this situation using a meta-predicate, *is-formula*,

is-formula(resolve,
        situation(Object, Dir),
        $F = M.A$,
        mass(Object, M) &
        accel-compt(Object, A, Dir) &
        sum-forces(Object, F, Dir)

The first parameter is the name of the physical formula. The second parameter is the situation in which the formula is being used, in this case the object being resolved about and the direction of resolution. The third parameter is the formula itself and the fourth parameter relates the second to the third by a series of propositions. When it has decided what formula to use and in what situation, MECHO can fill in the blanks in the formula – the F, M and A – by setting up the propositions in the fourth parameter place as subgoals, e.g. by trying to satisfy mass(Object, M), where the variable Object is already matched to some particular constant, say part1. There is already a proposition in the intermediate representation which will match mass(part1, M), but this is not so for accel-compt(part1, A, 270). There is a gap between the information provided in the problem statement and the information needed to solve the problem.

As already discussed, inference is used to bridge this gap. The first axiom MECHO will need is

accel(Object, Acc, Dir1) &
$A = \text{Acc.cos(Dir}-\text{Dir1)}$
    $\rightarrow$ accel-compt(Object, A, Dir)

and then the axiom relating the accelerations of the two particles in a pulley system discussed earlier.

To satisfy sum-forces(part1, F, 270) a series of inferences are made, in which the various kinds of forces which might be involved, e.g gravitational, reactions, friction, etc are considered in turn. The assumption is made that only those forces are acting which we can deduce are acting – that we have complete information about the situation.

14.5 Choosing Equations

We have seen how MECHO can form an equation from the intermediate representation, provided it knows what formula to use and what situation to use it in, but how can it make this choice?

The guiding principle is to choose equations which will help it to solve for the *sought* quantities in terms of the *given* ones. Suppose, for instance, it decides to form an equation which solves for the acceleration, acc. This quantity provides a lot of clues as to what formula and situation might be used. For instance, not all the formulae relate accelerations to other quantities. Those that do not can be ignored. acc is the acceleration of a particular object, part2, and this particle is moving in direction 90. This suggests the situation, situation(part2, 90). The equation chosen should be independent of any which have already been chosen to solve for other quantities. If any choices remain we would prefer an equation which does not introduce any new unknowns .

MECHO uses all this information when choosing an equation. Its procedure can be summarized as follows.

- Analyse the quantity currently being solved for. Find out what sort of quantity it is and what situation it is defined in, e.g. the acceleration in direction 90 of part2.

- On this basis select a particular formula and situation.

- Fill in the blanks in the formula using inference if necessary.

- If any of these inference steps would require the introduction of some new quantity, e.g. the velocity of part1, then remember this node as a continuation node, but go back and try another formula and/or situation.

- If an equation cannot be formed without introducing new quantities then introduce some by resurrecting a continuation node. Mark the new quantities as sought unknowns and solve for them later.

- Check that the new equation is independent from any that were formed earlier.

- Repeat the process on any remaining sought quantities.

- Use the measure propositions to translate all the quantities into numbers in compatible units.

The result of this procedure is a list of equations, e.g.

$$m1.g - t = m1.a$$
$$t - m2.g = m2.a$$

together with a record of which equations solve for which unknowns. MECHO feeds these to PRESS (see chapter 12) for solution.

14.6 Meta-Level Knowledge

In the last three sections we have seen how MECHO cues schemata, makes inferences and chooses equations. All these processes involve choices: choices which could cause combinatorial explosions unless they are controlled. To control them MECHO uses the technique of meta-level inference described in chapter 12. Using this technique necessitates the representation of meta-level knowledge about the expressions of the intermediate representation, physical formulae, etc.

We have already met one such piece of meta-level knowledge. The is-formula predicate is used to represent knowledge about the way that the variables of physical formulae are related to objects in the intermediate representation. Further examples are the predicate *kind* which relates a physical quantity to its type and its original defining proposition, e.g.

kind(acc, acceleration, accel(part2, acc, 90))

This information is used to analyse the quantity being solved for. The quantity type is used, in conjunction with the meta-predicate, *relates*, to draw up a short list of physical formulae which involve quantities of this type. That is,

relates(acceleration, <resolve, const-accel1, const-accel2,...>)

says that the formulae in the list, <resolve, ...> all involve accelerations. The defining proposition is further analysed by a complex process of meta-level inference to help suggest what situation to focus on when forming equations, e.g. the proposition accel(part2, acc, 90) will be broken apart to get part2 and 90. In an application of resolve, these might suggest resolving about part2 in direction 90 degrees, but the analysis will also range wider and suggest resolving about objects in contact with part2 etc.

To control the making of inferences, various kinds of meta-knowledge are used. One of the most powerful techniques uses the functionality of some of the predicates, i.e. the fact that some parameters of a predicate are functionally dependent on others. For instance, in mass(part1, m1), m1 is functionally dependent on part1, and hence m1 has the function properties of uniqueness and existence , i.e. every particle has exactly one mass (see section 4.3.1). MECHO uses this knowledge in three ways.

- *Uniqueness test*: before attempting to prove that mass(part1, m2), say, MECHO looks to see that it does not already know mass(part1, m1), where m1 ≠ m2. If it does then the original goal is abandoned as unsatisfiable.

- *Back-up Prevention*: if the goal mass(part1, M) is satisfied by {m1/M} then further choice points are deleted, since part1 can have no other mass.

- *Controlled Creation*: if the goal mass(part1, M) should fail then MECHO can choose to force satisfaction by creating a Skolem constant, say m, defined to be the mass of part1. The decision as to whether to do this depends on the state of the problem solving process, e.g. whether an attempt has already been made to form an equation not containing intermediate unknowns.

This incorporation of the uniqueness and existence properties of functions into the inference control mechanism, more than compensates for the loss of functional information implied by MECHO's use of predicates to represent functional relationships like the mass and acceleration of a particle. The extra power comes from MECHO's ability to use the functional knowledge to control what inferences are made and what terms are introduced into the Herbrand Universe.

14.7 Summary

We have seen that it is possible to model aspects of mathematical reasoning other than the proving of theorems, in particular, the formation of a mathematical model from an informal specification of a problem.

When this informal specification is in a natural language like English, a natural language understanding program is required. This translates from English to some more formal representation of the problem. In the case of very simple algebra word problems this formal representation can be sets of equations. But when the problem specification includes 'scene setting' information an intermediate representation is needed.

To bridge the gap between this intermediate representation and the knowledge required to form equations it is necessary to cue schemata and to make inferences. These inferential processes involve choices, so some control mechanism is required. The technique of meta-level inference, described in chapter 12, can be adapted to this task.

Further Reading Suggestions

[Bobrow 64] is a good introduction to STUDENT. [Bundy et al 79a] is a good introductory account of MECHO. [Bundy et al 79b] is a more detailed account.

Part V:
Technical Issues

15
Clausal Form

☐ In this chapter we investigate clausal form.
☐ Section 15.1 defines prenex normal form.
☐ Section 15.2 defines Skolem normal form.
☐ Section 15.3 defines a technique for Skolemizing non-prenex formulae.
☐ Section 15.4 defines conjunctive normal form.
☐ Section 15.5 completes the definition of clausal form and the description of a procedure for putting formulae into clausal form.
☐ Section 15.6 proves that a formula is weakly equivalent to its clausal form.
☐ Section 15.7 gives a neat test for the truth of a formula in clausal form. This test proves useful in the next chapter.

Clausal form is a concatenation of several of the traditional logic normal forms, namely *prenex normal* form, *Skolem normal form* and *conjunctive normal form*, applied in that order. We will consider each of these in turn, but before doing so we must perform some trivial preprocessing on the formulae.

The theorem proving processes, we will describe, work only on sentences, so we must ensure that both the axioms, Ax, and the conjecture, Thm, are fully closed with universal quantifiers applied to their free variables before proceeding further. This is particularly important for the conjecture, which is to be negated, because the operations of negation and closure are *not* commutative, e.g. $\forall X \sim X=0$ does not mean the same as $\sim \forall X \ X=0$, in fact the first is false in arith whereas the second is true.

The second piece of preprocessing is to use the equivalences of section 2.2.2 to eliminate the connectives $\rightarrow$ and $\longleftrightarrow$. We could also get rid of & (or alternatively v) at this stage, but we will find it useful in the sequel to have both v and &.

We now have a sentence containing only the connectives $\sim$, v and &. Notice that the processes of forming the closure of the original formula and eliminating the connectives $\rightarrow$ and $\longleftrightarrow$, both preserved the meaning of the formula, i.e. the resulting formulae were equivalent to the original one.

15.1 Prenex Normal Form

The idea of prenex normal form is to move all the quantifiers to the top of the sentence, from whence they will be removed in the 'Skolemization' step of the next section.

We will define the normal form by giving a set of *rewrite rules,* that is a set of rules for converting any sentence into a sentence in normal form by stages. Each rule deals with a sentence which is not in normal form, because it contains a quantifier not at the top and shows how to move the quantifier to the top. A quantifier which is not at the top must have another symbol immediately above it: in this case the symbol will be one of the connectives, ~, & or v. Since the quantifier can be in either parameter position of & or v, and there are two quantifiers, there are 10 cases to consider. The rules for each of these cases is given in table 15-1.

Table 15-1: Rewrite Rules for Prenex Normal Form

Before		After
~ ∀X A	=>	∃X ~A
~ ∃X A	=>	∀X ~A
∀X (A) & B	=>	∀X (A & B)*
∃X (A) & B	=>	∃X (A & B)
A & ∀X (B)	=>	∀X (A & B)
A & ∃X (B)	=>	∃X (A & B)
∀X (A) v B	=>	∀X (A v B)
∃X (A) v B	=>	∃X (A v B)
A v ∀X (B)	=>	∀X (A v B)
A v ∃X (B)	=>	∃X (A v B)

Exercise 45: If we had not eliminated the → connective in terms of ~ and v before we started we would also have had to give the 4 rewrite rules for → in table 15-1. (This illustrates the time saving benefits of a normal form.) Work out what these rules would have been by using the rules for ~ and v. Are you surprised at the result? Can you give 4 similar rules for ←→? If not, why not?

To put a sentence in normal form apply the following procedure.

(a)If the sentence is already in normal form then stop with success.

(b)Otherwise it contains a subformula consisting of a connective immediately above a quantifier.

(c)Pick such a subformula and find which of the expressions in the left hand side of table 15-1 matches it.

(d)Replace the subformula by the expression on the right hand side of the table.

*In case B already contains X, it is necessary first to substitute a new variable for X in ∀X (A) to prevent any confusion. Similar remarks hold for the remaining rules.

Lemma 1: The prenex normal form process, applied to any sentence, terminates producing an equivalent sentence in prenex normal form.

To see that this is so consider first the termination. Let dpth be the total depth of all the quantifiers, i.e. the sum of all the distances from the root of the sentence tree to a quantifier. Each time a rule is applied dpth will be reduced by 1. dpth can never be negative so the process must terminate with the sentence in normal form.

The sentence output by this process is equivalent to the one input because each application of a rule preserves equivalence. Consider, for instance, the first rule. Now

$\sim \forall X\ A(X)$ is true in an interpretation I
        iff
$\forall X\ A(X)$ is false in I
        iff
for some c in the universe of I, $A(c)$ is false in I
        iff
$\sim A(c)$ is true in I
        iff
$\exists X \sim A(X)$ is true in I.

We can establish the same truth preserving properties for each of the rules in table 15-1

Exercise 46: Establish the equivalence properties of two more of the rules from table 15-1.

15.2 Skolem Normal Form

In Skolem normal form we remove all the quantifiers, both universal and existential, leaving a formula with only free variables. This process is made easier by having the original formula in prenex normal form, and we assume this in this section. In the next section we show how to Skolemize any formula.   Removing the universal quantifiers while preserving meaning is easy since a formula and its closure mean the same thing. We need only omit the quantifiers. Doing the same for the existential quantifiers is harder. The key idea is to introduce a new function to replace the bound variable. These are called *Skolem functions*. When the function are nullary they are called *Skolem constants*.

Consider the sentence

$$\exists X \quad X^2 + 2.X + 1 = 0 \tag{i}$$

This asserts the existence of a number satisfying a quadratic equation. The same effect could be achieved by asserting the equation with a particular number substituted for X. But since nothing more is asserted by (i) than that the number satisfies the equation, it would not do to substitute a previously known number for X, e.g. 3 or −1. We will have to invent a new number, the Skolem constant x, just for the purpose.

$$x^2 + 2.x + 1 = 0$$

We might then use our knowledge of algebra to prove that

$$x = -1$$

but this is beyond the scope of our present concerns.

If the existential quantifier in question is dominated by universal quantifiers, for instance,

$$\forall A \, \forall B \, \forall C \, \exists X \quad A.X^2 + B.X + C = 0 \tag{ii}$$

then we will have to use a Skolem function rather than a constant. The parameters to the function will be universally quantified variables, thus allowing a different value for each different combination of universal variables.

$$A.x'(A,B,C)^2 + B.x'(A,B,C) + C = 0$$

Again we could use our knowledge of algebra to prove that

$$x'(A,B,C) = (-B \pm \sqrt{B^2 - 4.A.C})/2.A$$

but such proofs will not always be available to us.

Exercise 47: Skolemize the sentence

$$\forall M \, \exists Delta \, \forall X \quad |X| \leqslant Delta \rightarrow 1/X > M$$

Note that $\forall M$ dominates $\exists Delta$, but that $\forall X$ does not.

When putting a formula into Skolem normal form it is necessary to keep a record of the universally quantified variables which dominate each existentially quantified one. Thus the rewrite rule table requires an extra 'action' column to say what must be done to use and update this record each time a rewrite occurs.

Table 15-2: Rewrite Rules for Skolem Normal Form

Before	After	Action
$\forall X \, A(X)$	=> $A(X)$	Add X to the set of free variables.
$\exists X \, A(X)$	=> $A(x(V))$	Where V is the vector of free variables and x is a new Skolem function.

These rules are always applied to the whole sentence and thus peel off the quantifiers, topmost first, until they are all eliminated.

Lemma 2: The Skolemization process, applied to a sentence in prenex normal form, terminates with a weakly equivalent formula in Skolem normal form.

To see that it terminates is easy, since the number of quantifiers is reduced on each application. The rules continue to apply while a quantifier remains at the top of the formula, and all formulae are in prenex form, with their quantifiers at the top, so the final formula will have no quantifiers, i.e. it will be in Skolem normal form.

The input and output sentences are not necessarily equivalent, but their meanings are closely related, and we have captured this relationship with the *weakly equivalent* term, which we will now explain.

It is easy to see that the application of the first rule, which eliminates universal quantifiers, also preserves equivalence. The difficulties arise with the application of the second rule: the conversion from

$$\exists X\ A(X) \text{ to } A(x(V))$$

Now

$A(x(V))$ is true in I.
      implies
for all substitutions of objects of the universe v for V
$A(x(v))$ is true in I.
      implies
for all substitutions of objects of the universe v for V
there is a substitution of an object of the universe for X such that
$A(X)$ is true in I.
      implies
$\exists X\ A(X)$ is true in I.

So the original formula is a logical consequence of the final one, but the argument does not reverse, because the object of the universe, t, which, when substituted for X, makes $A(X)$ true, may not be $x(v)$. However, since x is a newly created Skolem function it will not appear in $A(X)$, nor in any existing mathematical theories, so it will not be assigned a value or calculation procedure in any interpretation of $A(X)$. Thus any model of $\exists X\ A(X)$ can be extended to be a model of $A(x(V))$ by assigning a value or calculation procedure to x such that

$$x(v) = t$$

This is what we mean by weakly equivalent, and we will see that it is sufficient for our purposes.

15.3 Skolemizing Non-Prenex Formulae

Although Skolemization is easy to explain and justify on sentences in prenex normal form, it can actually be applied directly to any formula. The process involved is not easily explained using rewrite rules. Instead we will use a set of equations defining a function *skolem* with two parameters: the formula to be Skolemized and the set of free variables it contains, and whose result is the Skolemized formula, e.g.,

$$\text{skolem}(\forall X\ p(X), \{\}) = \text{skolem}(p(X), \{X\}) = p(X)$$

When the topmost connective is v or & the situation is simple: skolem is merely called recursively on each of the arguments of the connective. To see this, consider, for instance,

$$\text{skolem}(P\ v\ \forall x\ Q(x), \{\})$$

If we first put the formula in prenex form

$$\text{skolem}(\forall x\ (P\ v\ Q(x)), \{\})$$

then Skolemize it

$$P\ v\ Q(x)$$

the effect is the same as if we had Skolemized the arguments

$$\text{skolem}(P, \{\})\ v\ \text{skolem}(\forall x\ Q(x), \{\})$$
$$= P\ v\ Q(x)$$

When the topmost connective is $\sim$ the situation is a little more complicated. Consider

$$\text{skolem}(\sim \exists x\ P(x), \{\})$$

If this formula is put in prenex form

$$\text{skolem}(\forall x\ \sim P(x), \{\})$$

the quantifier changes its type, from existential to universal. The effect

$$\sim P(x)$$

is as if a dual Skolemization were applied to the argument of $\sim$ in which quantifiers were treated as if their types were reversed.

$$\text{skolem}(\sim \exists x\ P(x), \{\})$$
$$= \sim \text{dual-skolem}(\exists x\ P(x), \{\})$$
$$= \sim P(x)$$

With *dual-skolem* acting on an existential quantifier as if it were a universal one. Of course, if dual-skolem goes inside another $\sim$ then it turns back into skolem again.

The simplest way to handle this is to define a new version of skolem, called *new-skolem* with an extra parity argument, either 'regular' or 'dual'. If the argument is regular then new-skolem behaves like skolem. If the argument is dual then new-skolem behaves like dual-skolem. *opposite* will be a new function which returns dual given regular and vice versa.

skolem(A,V) = new-skolem(A,V,regular)
dual-skolem(A,V) = new-skolem(A,V,dual)

opposite(regular) = dual
opposite(dual)=regular

The equations for new-skolem are then

new-skolem($\forall$x A(x), Vars, Par)
   = new-skolem($\exists$x A(x), Vars, opposite(Par))

new-skolem($\exists$x $\sim$ A(x), $\{y_1,...,y_n\}$, regular)
   = new-skolem(A(f($y_1,...,y_n$)), $\{y_1,...,y_n\}$, regular)
where f is a new Skolem function

new-skolem($\exists$x A(x), Vars, dual)
   = new-skolem(A(x), $\{x\}$ U Vars, dual)

new-skolem($\sim$ A, Vars, Par)
   = $\sim$ new-skolem(A, Vars, opposite(Par))

new-skolem(AvB, Vars, Par)
   = new-skolem(A, Vars, Par) v new-skolem(B, Vars, Par)

new-skolem(A&B, Vars, Par)
   = new-skolem(A, Vars, Par) & new-skolem(B, Vars, Par)

With the aid of the equations for $\sim$ and v we can define the effect of Skolemization on A$\rightarrow$B.

new-skolem(A$\rightarrow$B, Vars, Par)
   = new-skolem($\sim$A v B, Vars, Par)
   = new-skolem($\sim$A, Vars, Par) v new-skolem(B, Vars, Par)
   = $\sim$ new-skolem(A, Vars, opposite(Par)) v
     new-skolem(B, Vars, Par)

$\quad$ = new-skolem(A, Vars, opposite(Par)) $\rightarrow$
$\quad$ new-skolem(B, Vars, Par)

Exercise 48: Try this same technique on A$\longleftrightarrow$B. What happens?

15.4 Conjunctive Normal Form

The idea of conjunctive normal form is that each type of connective appears at a distinct height in the formula. The negation signs, $\sim$, are lowest, with only propositions as their parameters. The disjunction signs, v, are in the middle, with only propositions or negated propositions as their parameters. And the conjunction signs, &, are at the top, taking the disjunctions as their parameters. For instance,

$$\{p \vee \sim q \vee \sim r\} \ \& \ \{\sim p \vee \sim q\} \ \& \ \{\sim p \vee r\}$$

The negated or unnegated propositions are called literals. The disjunctions of literal are called clauses.

$\quad$ We will transform formulae into conjunctive normal form in two stages. The first of these will move the negation signs down, until they are next to the propositions. Let us call the resulting intermediate normal form *literal normal form*, since negation signs will only appear in literals. We need only consider the cases where a $\sim$ appears at the top most level, dominating another connective and show how it is to be removed or sent to a lower level. There are three of these cases to consider: where a $\sim$ dominates another $\sim$, another v or another &. The transformations for these cases are given in table 15-3.

Table 15-3: Rewrite Rules for Literal Normal Form

Before		After
$\sim\sim$A	=>	A
$\sim$(A v B)	=>	$\sim$A & $\sim$B
$\sim$(A & B)	=>	$\sim$A v $\sim$B

$\quad$ These rules are to be applied in a similar way to the rules for prenex normal form, taking any subformula where a $\sim$ dominates an & or v and applying a rule to it until no further rules apply.

$\quad$ *Lemma 3*: The literal normal form process, applied to a formula in Skolem normal form, terminates producing an equivalent formula in literal normal form.
$\quad$ *Proof*: To see that the process terminates consider the following numeric function on formulae. Let the *weight* of a $\sim$ sign be the size of the formula it dominates. (Consider the formula as a tree: the weight of

~ is the number of arcs below it.) Let the *load* of a formula be the sum of the weights of all the ~ signs in it. The load of a formula will reduce each time a rule is applied. Since it cannot reduce to less than 0 then there must come a time when no further rule applies. To see that the load reduces on each rule application, consider each rule in turn.

– If the first rule is applied then the load is reduced by the weights of two ~ signs.
– If the second or third rule is applied then the load is reduced by the weight of one ~ sign, but increased by the weights of two new ones. However, the combined weights of the new ~ signs is 2 less than the weight of the old one. Hence there is a net reduction of 2.

By the usual argument the process cannot terminate unless the formula is in literal normal form; and the input and output formula are equivalent because each rule preserves equivalence. To see this in the case of the second rule, consider the semantic trees for the two formulae involved. These are given in figure 15-1.

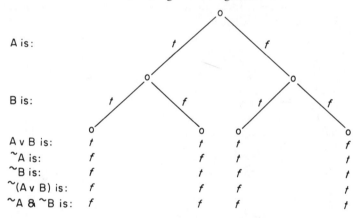

A is:

B is:

A v B is:
~A is:
~B is:
~(A v B) is:
~A & ~B is:

*Figure 15-1:*The Equivalence of Two Literal Normal Form Rules

Similar arguments hold for the other two rules.   QED

Exercise 49: Establish the equivalence property of the third rule for literal normal form.

We now have a formula like

$$\{(p \ \& \ q) \ v \sim r\} \ \& \ p$$

in which ~s only appear at height 1, if at all. It only remains to shift the &s to the top and vs below them. This can be done with only two rules (and these are essentially variants of a single rule).

Table 15-4: Rewrite Rules for Conjunctive Normal Form

Before		After
A v (B & C)	=>	(A v B) & (A v C)
(B & C) v A	=>	(B v A) & (C v A)

This is because the only cases to consider are where a v dominates an & symbol. Applied whenever a v appears immediately above an &, these rules will gradually decrease the maximum depth of the &s and decrease the maximum height of the vs, until all the &s are above all the vs, e.g.

$$\{(\sim p \ \& \ q) \ v \sim r\} \ \& \ p \longleftrightarrow (\sim p \ v \sim r) \ \& \ (q \ v \sim r) \ \& \ p$$

By the usual arguments

> *Lemma 4:* The conjunctive normal form process, applied to a formula in literal normal form, terminates producing an equivalent formula in conjunctive normal form.

> *Exercise 50:* Put the formula

> $$\{(\sim p \ v \ q) \ \& \sim r\} \ v \ p$$

in conjunctive normal form.

> *Exercise 51:* Check that the first pair of entries in table 15-4 are equivalent formulae

> *Exercise 52:* Using the rewrite rules

> A & (B v C) => (A & B) v (A & C)
> (B v C) & A => (B & A) v (C & A)

which are the duals of those in table 15-4, put the formula in exercise 50 in disjunctive normal form, in which the formula is a disjunction of a conjunction of literals.

15.5 Clausal Form

Our last transformation will be to break up the conjunction of clauses into the set of individual clauses and to standardize the variables apart by ensuring that no variable appears in more than one clause. In our running example we will break up

$\{p(X,Y) \lor \sim q(X) \lor r(Z)\}$ & $\{\sim p(X,Y) \lor \sim q(X)\}$ & $\{\sim p(X,Y) \lor r(Z)\}$

into the set

$p(X,Y) \lor \sim q(X) \lor r(Z)$
$\sim p(X,Y) \lor \sim q(X)$ and
$\sim p(X,Y) \lor r(Z)$

and then standardize apart to get

$p(X1,Y1) \lor \sim q(X1) \lor \sim r(Z1)$
$\sim p(X2,Y2) \lor \sim q(X2)$ and
$\sim p(X3,Y3) \lor r(Z3)$

where each clause has different variables.

The first part of this transformation relies on the rewrite rules

A & B => A
A & B => B

applied repeatedly to split the conjunction into its individual clauses.

The second part uses the substitution rule

$$\frac{A(X)}{A(T)}$$

to replace the variables in each clause with new variables in such a way that no two clauses share a common variable.

Our formulae are now in *clausal form*, and this is the normal form required by the resolution rule. It is clear that

Lemma 5: The clausal form process, applied to a formula in conjunctive normal form, terminates producing a formula in clausal form.

Clausal form might seem a highly constrained and unnatural representation for mathematical knowledge, and it has been criticised on those grounds, but we will see that it is in fact surprisingly natural in many cases. In fact most of the axioms given in section 4.2 and many of those to be given in the future are expressed in clausal form, provided we extend that notion to include Kowalski form.

To put a clause into Kowlaski form gather all the negation signs in each clause into one sign, using the rule

$\sim A \lor \sim B => \sim(A$ & $B)$

repeatedly. Now eliminate this one negation sign completely with a single application of

$\sim A \lor B => A{\rightarrow}B$

Thus p v ~q v ~r will be transformed first to p v ~(q & r) and then to q & r → p.

15.6 Weak Equivalence

If A is a sentence then let nf(A) denote the formula obtained by the repeated application of: prenex normal form; Skolem normal form; literal normal form; conjunctive normal form and clausal form. From lemmas 1, 2, 3, 4 and 5 it is clear that nf(A) is weakly equivalent to A. From this we can deduce the theorem.

Theorem 6: A is unsatisfiable iff nf(A) is unsatisfiable.

Proof: Only If
Suppose that A is unsatisfiable, but that nf(A) has a model, M. Since A is a logical consequence of nf(A), M is a model of A, contradicting the unsatisfiability of A. Hence nf(A) is also unsatisfiable.
     If
Suppose that nf(A) is unsatisfiable, but that A has a model, M. M can be extended to a model M' of nf(A) by associating suitable calculation procedures with the Skolem functions of nf(A). This contradicts the unsatisfiability of nf(A). Hence A is also unsatisfiable.   QED

15.7 The Meaning of Formulae in Conjunctive Normal Form

There is a particularly neat characterization of the meaning of formulae in conjunctive normal form.

Lemma 7: A formula in conjunctive form is unsatisfiable iff for each interpretation, I, there is an instance of the formula such that the negations of all the literals in one of the clauses are true in I.

Proof: If the formula is unsatisfiable then it is not true in every interpretation, i.e. for each interpretation there is an instance which is false.

Let

C_1 &...& C_n where C_i is a clause

be the false, instance for the interpretation I. From the semantic tree for & we can deduce that one of the C_i must be false in I. Suppose this false clause is

L_1 v...v L_m where each L_j is a literal

then by the semantic tree for v all of the L_j are false in I, and by the semantic tree for ~ all of the L_j are true in I.

This argument is reversible.

This lemma will be invaluable in the proofs of Herbrand's theorem and the completeness of resolution.

15.8 Summary

We have seen that any predicate calculus formula can be converted into a weakly equivalent formula in either conjunctive normal form, clausal form or Kowalski form. The first of these normal forms is useful for proving theoretical results about resolution. The second is useful for defining resolution and paramodulation. The third is useful for making natural looking formal proofs. We have introduced the notion of weak equivalence between two formulae; that one has a model iff the other does. This will be just the concept of equivalence used in the theoretical results on resolution of the next chapter.

16

Herbrand Proof Procedures

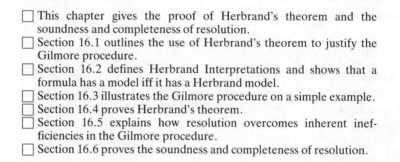

☐ This chapter gives the proof of Herbrand's theorem and the soundness and completeness of resolution.

☐ Section 16.1 outlines the use of Herbrand's theorem to justify the Gilmore procedure.

☐ Section 16.2 defines Herbrand Interpretations and shows that a formula has a model iff it has a Herbrand model.

☐ Section 16.3 illustrates the Gilmore procedure on a simple example.

☐ Section 16.4 proves Herbrand's theorem.

☐ Section 16.5 explains how resolution overcomes inherent inefficiencies in the Gilmore procedure.

☐ Section 16.6 proves the soundness and completeness of resolution.

In this chapter we will prove Herbrand's theorem and the soundness and completeness of resolution. These proofs provide the theoretical justification for two proof procedures: the *Gilmore procedure* and the *Resolution procedure*. The Gilmore procedure is introduced because it is a direct application of Herbrand's theorem, and serves as a useful introduction to the Resolution procedure. Our principal concern is to show that by restricting ourselves: to a refutation system, with a single rule of inference, resolution, applied only to formulae in clausal form, we can still prove all correct argument forms.

16.1 The Significance of Herbrand's Theorem

In order to explain the Gilmore procedure we will prove Herbrand's Theorem [Herbrand 30] and discuss how it can be used as the basis of a computer program for showing that a formula is a logical consequence of a set of axioms. The theorem is:

> A formula, A, in conjunctive normal form is unsatisfiable iff there exists a contradiction consisting of a finite conjunction, A', of instances of clauses of A.

but before we prove it let us investigate its significance.

Suppose we have some axioms, Ax, and we want to see if some conjecture, Thm, is a logical consequence of Ax. The key idea is to add the negation of Thm to Ax and show that the resulting set of axioms is unsatisfiable, by putting it into clausal form and finding a contradictory, finite conjunction of clause instances. The theorems we shall prove in this chapter will enable us to show that

Thm is a logical consequence of Ax
iff
Thm' is a logical consequence of Ax', where Thm' and Ax'
are the closures of Thm and Ax, respectively
iff
Ax' & ~Thm' is unsatisfiable
iff
S, the clausal form of Ax' & ~Thm' is unsatisfiable
iff
There exists a contradiction
consisting of a finite conjunction of instances of clauses of S.

This suggests a way of testing to see if Thm is a logical consequence of Ax, namely

– Put Ax & ~Thm into clausal form. Let S be the result.

– Generate instances of clauses of S.

– Test conjunctions of these clauses for contradiction using semantic trees as described in section 3.

As mentioned in section 1.3.3 this is essentially the process suggested by Gilmore in 1960 and implemented by him in a computer program [Gilmore 60].

As it stands the Gilmore procedure is not a very practical test, because there are too many ways to form instances. We must consider the universe of every possible interpretation and use the objects of each universe in all possible ways. We will see in the next section that we can improve on this considerably because there is a class of interpretations, called *Herbrand Interpretations*, which can stand for all the others. The Herbrand Interpretations share the same universe, called the Herbrand Universe. Hence we need only consider instances made by substituting elements of the Herbrand Universe for variables.

With a little care we can make sure that none of the possible conjunctions of instances of clauses is omitted from our testing. Then we can be sure that if Thm is a logical consequence, we will find out eventually. However, the search through possible instances could go on forever. In fact the following cases can arise.

1. Thm is a logical consequence of Ax. In this case our search will eventually terminate with success, i.e. we will find a contradiction.

2. Thm is not a logical consequence. There are two possible sub-cases.

   (a) We run out of instances to generate, without having found a contradiction. In this case we can be sure that Thm is not a logical consequence of Ax.

   (b) We go on generating instances for ever and never find a contradiction. In this case we are never sure whether Thm is a logical consequence of Ax or not. We could be in either case 1 or 2.

16.2 Herbrand Interpretations

In this section we define the Herbrand Interpretations of a formula, and we prove that these are the only interpretations that we need to consider when testing the unsatisfiability of a formula. We start by defining the Herbrand Universe of a formula. What must this universe contain? For each n-ary function, f, in the formula and n objects, $t_1,...t_n$, in the universe it must contain the result of applying f to the t_i s. This can be achieved by a simple syntactic device; we put all the variable free terms in the universe and let each term denote the result of applying its function to its parameters. We will call this syntactic device, *self denotation*.

> *Definition 1:* The Herbrand Universe of a formula consists of all the variable free terms that we can make from the constants and functions in the formula.

In this section it is useful to distinguish the constants from the other functions, so by 'function' we mean non-nullary function.

This definition applied to formula

$$\forall X \, \exists Y \;\; X = Y \lor X = Y + 1$$

for instance, yields the function + and the constant 1. If we were considering this formula in the context of the axioms of some theory then these axioms would provide additional functions and constants, but we ignore this possibility for the sake of simplicity. The constants thus obtained are our first batch of variable free terms. But these are not enough. We must consider all possible ways in which new terms may be built up by combining old terms with functions, e.g.

$$\{1, 1+1, (1+1)+1, 1+(1+1), (1+1)+(1+1), \ldots\ldots, \}$$

Three cases may arise

1. If the axioms contain both constants and functions then the Herbrand Universe will always be countably infinite.

2. If the axioms contain only constants then the Herbrand Universe will be the finite set of these.

3. If the axioms contain no constants then to prevent having an empty Universe we will add one, say a. This puts us back in either case 1 or 2 above.

The only way in which Herbrand Interpretations differ is over assignment of predicate meanings. A convenient way of assigning these meanings is to consider the truth value of each variable free instance of each proposition in the formula. These instances can be obtained by forming the set of propositions in the formula, and then substituting terms from its Herbrand Universe for its variables in all possible ways. We call this set the *Herbrand Base* of the formula. For instance, the Herbrand Base of

$$\forall X \, \exists Y \;\; X = Y \lor X = Y + 1$$

is formed by taking its set of propositions $\{X = Y, \; X = Y + 1\}$ and substituting the terms,

$$\{1, 1+1, (1+1)+1, 1+(1+1), (1+1)+(1+1), \ldots\ldots, \}$$

for X and Y, in all possible ways, to form

$$\{1 = 1, \; 1 = 1 + 1, \; 1 + 1 = 1, \; 1 + 1 = 1 + 1, \ldots.\}$$

Note that the Herbrand Base can always be enumerated by a finite or infinite list. In particular, if a formula contains no functions then its

Herbrand Universe will be finite and, hence, its Herbrand Base will be finite.

It is not necessarily the case that every proposition formed by applying predicates from the formula to terms from its Herbrand Universe is in the Herbrand Base. For instance, the proposition 1=1 is not in the Herbrand Base of ∀X ∃Y X=Y+1, because it is not an instance of X=Y+1. However the truth of the formula does not depend on these missing propositions.

Since the missing propositions do not contribute to the truth of a formula, each of its Herbrand Interpretations is effectively determined by a mapping from the Herbrand Base to the set of truth values, $\{t,f\}$. For instance, the mapping

$$
\begin{array}{lcc}
1=1 & \mapsto & t \\
1=1+1 & \mapsto & f \\
1+1=1 & \mapsto & f \\
1+1=1+1 & \mapsto & t \\
1=(1+1)+1 & \mapsto & t \\
1+1=(1+1)+1 & \mapsto & f \\
\ldots\ldots \text{etc} & &
\end{array}
$$

defines a Herbrand Interpretation which we will call *boole2*.

The truth of a formula can now be determined from the meaning rules for the quantifiers and the semantic trees for the connectives, as in section 3.3.2. However, it is not necessary to appeal to calculation procedures for the functions and predicates. Once the problem has been reduced to the meaning of variable free propositions then the Herbrand Base can be used to find their truth values, e.g.

∀X ∃Y X=Y ∨ X=Y+1 is true in boole2
   iff   (by meaning of ∀)
∃Y 1=Y ∨ 1=Y+1 is true in boole2 and
∃Y 1+1=Y ∨ 1+1=Y+1 is true in boole2 and
   .... etc
   iff   (by meaning of ∃)
either 1=1 ∨ 1=1+1 or 1=1+1 ∨ 1=(1+1)+1 or ... is true in boole2 and
either 1+1=1 ∨ 1+1=1+1 or 1+1=1+1 ∨ ... is true in boole2 and
   .... etc
   iff   (by semantic tree for ∨)
either 1=1, 1=1+1, 1=1+1, 1=(1+1)+1, ... is true in boole2 and

either 1+1=1, 1+1=1+1, 1+1=1+1, 1+1=(1+1)+1,...
is true in boole2 and

    .... etc
       iff   (by Herbrand Base)
either $t, f, f, t, ...$ is true in boole2 and
either $f, t, t, f, ...$ is true in boole2 and

    .... etc
       iff   (by meaning of English)

t

Since each Herbrand Interpretation of a formula is determined by a mapping of the Herbrand Base, we can encapsulate the set of all Herbrand Interpretations by extending our notion of semantic tree. If $p_1,..., p_n,...$ is an enumeration of the Herbrand Base of the formula, then each branch of the tree in figure 16-1 defines a mapping from the Herbrand Base to $\{t,f\}$, i.e. an interpretation. Of course, if the Herbrand Base is infinite the tree will be infinitely deep.

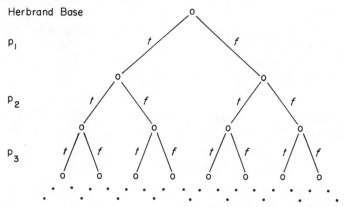

Figure 16-1: All Herbrand Interpretations of a Formula

To see that we can limit our attention to Herbrand Interpretations we must prove the lemma:

Lemma 2: A formula has a model iff it has a Herbrand Model.

Proof: The if part of this is demonstrated by showing that a Herbrand Model of a formula, A, really is a model of A. Since the Herbrand Model defines a universe and assigns a meaning in this universe to the functions using self denotation, it only remains to show that the mapping of the Herbrand Base assigns a meaning to the predicates, and that A is true under this assignment.

For each n-ary predicate, p, in A and each set of n terms, $t_1,...,t_n$ in its Herbrand Universe, we must assign a truth value to $p(t_1,..,_nt)$. If this proposition is in the Herbrand Base then we assign it the same truth value as is assigned by the Herbrand Interpretation. If it is not in the Herbrand Base then it does not matter what value we assign it, so let it be t. We must show that A is true in this interpretation, but this is trivial since, in either this interpretation or the Herbrand Model, A is true iff all its instances are true, and this is determined solely by those variable free propositions which are in the Herbrand Base.

To prove the only if part of the lemma we must show that if a formula, A, is true in an interpretation, I, then there is an Herbrand Interpretation HI in which A is true. Clearly HI will be constructed from I.

Let I have universe, U. If f is an n-ary function or constant in A, p is an n-ary predicate in A, and $a_1,...,a_n$ are n objects in U, let $u[f(a_1,...,a_n)]$ denote the member of U obtained by applying f to the a_is, and let $b[p(a_1,...,a_n)]$ denote the truth value obtained by applying p to the a_is.

Let HU be the Herbrand Universe of A and HB be its Herbrand Base. To define HI we must say which of the members of HB are true and which false.

Consider first the case when A contains some constants. Note that all these constants are members of U. We can define a function, huu, from objects in HU to objects in U, recursively as follows.

$huu(c) = c$ where c is a constant (and hence in U)
$huu[f(t_1,...,t_n)] = u[f(huu(t_1),...,huu(t_n))]$
                where f is an n-ary function
                and the t_i are variable free terms.

If A contains no constants then HU consists of all variable free terms built from the new constant, a, and the functions (if any) of A. U is non-empty, so let b be a member of U. We can define huu for this case recursively as follows:

$huu(a) = b$
$huu[f(t_1,...,t_n)] = u[f(huu(t_1),...,huu(t_n))]$
                where f is an n-ary function
                and the t are variable free terms.

huu can be used to define, hb, a function of propositions from HB to $\{t,f\}$ as follows

$hb[p(t_1,...,t_n)] = b[p(huu(t_1),...,huu(t_n))]$
                where p is n-ary predicate
                and the t_i are variable free terms.

Let the function hb define HI. We must now show that HI is a model of A.

- Let φ be a substitution of objects of HU for variables in A.

- Let φ' be the corresponding substitution of objects of U for variables in A obtained by replacing each pair t/X by huu(t)/X.

- A is true in I

- Therefore, each Aφ' is true in I.

- Therefore, each clause of Aφ' contains a literal, Lφ' which is true in I.

- But, by the definition of hb, Lφ' is true in I iff Lφ is true in HI.

- Hence each clause of Ao contains a literal, Lφ which is true in I.

- Hence, each Aφ is true in HI.

- Hence, A is true in HI.   QED

From now on, whenever we say interpretation we will mean Herbrand Interpretation. Similarly, the terms satisfiable, unsatisfiable, logically valid, etc will refer to Herbrand Interpretations.

Note that if a formula contains no functions then it has a finite Herbrand Universe, a finite Herbrand Base and only a finite number of Herbrand Interpretations. To test the logical validity of such a formula we test for truth in each interpretation. To test that a formula is true in an interpretation we need only test the truth of a finite number of variable free, quantifier free formulae and each of these can be tested using the tautology testing technique of section 2.2.4. Thus we have a decision procedure for theories without functions. This observation redeems a pledge made in section 4.3.2.

16.3 A Worked Example

Consider how the Gilmore procedure might work in a very simple example. Let Ax be the single axiom ∀X X=X and Thm be the sentence ∀Y Y=Y. To show that Thm is a logical consequence of Ax we put Ax & ~Thm into clausal form as follows:

∀X X=X & ~∀Y Y=Y
Skolemization replaces Y with Skolem constant y to get
X=X & ~y=y
which gives the two clauses
X=X and ~y=y

We then generate variable free instances of these clauses using the

Herbrand Universe, {y} (since y is the only constant and there are no proper functions). There are only two variable free instances,

y=y and ~y=y.

The conjunction of these is a contradiction, because no matter whether y=y is mapped to t or f, y=y & ¬y=y is f (see semantic tree for &).

The Gilmore procedure could prove that simple theorems were logical consequences of their axioms, i.e. theorems just a little harder than the trivial example above, but on non-trivial theorems it rapidly became engrossed in the endless possibilities for generating new instances. It could not see the wood for the trees!

However, Herbrand's Theorem and the Gilmore procedure served as the basis for the more able artificial mathematicians which followed. Herbrand's theorem is proved in the next section.

16.4 The Proof of Herbrand's Theorem

The idea behind the proof is that if a formula is unsatisfiable then for every Herbrand Interpretation there is an instance that is false in that interpretation. We would like to conjoin all these false instances to get a contradiction (which would of course be variable free), but since there might be infinitely many models this conjunction might be infinite. The proof shows how a finite conjunction can be made by having sets of Herbrand Interpretations share a single false instance. This is possible because the false instance for each interpretation is already false in a finite sub-interpretation and this sub-interpretation is a proper subset of a number of different interpretations.

We can represent all these sub-interpretations as a subtree of the complete semantic tree for a formula, called a *failure tree*.

> *Definition 3:* The failure tree of a formula, A, in conjunctive normal form is a minimal finite subtree of its semantic tree such that for each branch, B, there is a clause of A which is not true in the sub-interpretation defined by B.

The technique described above can now be caught in the following lemma, which will be used in the heart of the proofs of both Herbrand's Theorem and the Resolution Completeness Theorem.

> *Lemma 4:* If a formula, A, in conjunctive normal form is unsatisfiable then it has a failure tree

> *Proof:* Let I be the interpretation defined by a typical branch, B, of the semantic tree for A.

Assume A is unsatisfiable then by lemma 7 there is an instance, Aφ such that the negations of all the literals, Lφ in one of the clauses Cφ are mapped to true by I.

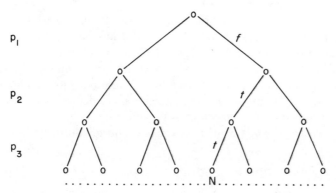

p_1

f

p_2

t

p_3

t

. . . . N . . . .

Figure 16-2: Cφ ≡ p_1 v ~p_3 fails at N

But since there are only finitely many literals, Lφ in Cφ there exists a smallest subbranch, B' of B, such that the falseness of Cφ is already determined by the sub-interpretation I' defined by B', see figure 16-2 for example: the truth values assigned to p_1 and p_3 on any branch passing through N are sufficient to ensure that p_1 v ~p_3 is false. Hence C is not true in I'.

The tip N is called a failure node for C and Cφ

The branches B' define a subtree T' of T. Since each B' is finite then T' is finite. If necessary, prune any nodes from T' which are ancestors of failure nodes to form T" (a minimal tree). T" is a failure tree for A.   QED

Herbrand's Theorem is now a simple consequence of this lemma.

Theorem 5: Herbrand's Theorem

A formula, A, in conjunctive normal form is unsatisfiable iff there exists a contradiction consisting of a finite conjunction, A', of instances of clauses of A.

Proof: If (the easy direction)

Let I be an interpretation defined by a typical branch of the semantic tree for A.

Assume A' is contradictory then A' is false in I. So some clause Cφ of A' is false in I (by semantic tree for &), where Cφ is an instance of a clause C in A.

So C is not true in I and hence A is not true in I. But I was a typical interpretation, so A is unsatisfiable.

Only If (the hard direction)

Assume A is unsatisfiable then by lemma 4 it has a failure tree, T".
Let B" be typical branch of T" and I" the sub-interpretation it defines.
From the definition of a failure tree there is an instance, Cφ of a clause,
C, of A which is false in I".

Let A' be the conjunction of all such Cφ. By the semantic tree for &,
A' is false in I". T" is finite and hence has only finitely many tips.
There is one such Cφ for each tip of T", so A' is finite.

T" is a semantic tree for A'. (Well not quite, but clearly we could
massage T" into a semantic tree for A', removing propositions not in
A' and extending all branches to the same depth. All the essential
properties would be inherited by the new tree.)

A' is false at every tip of T", so A' is a contradiction.   QED

16.5 The Resolution Procedure

How can the Gilmore procedure be improved? What are its inherent
inefficiencies?

– Many of the possible conjunctions of instances of clauses from nf(Ax
   & ~Thm) are isomorphic. Thus work in testing for contradiction is
   essentially duplicated. The failure to find a contradiction in one
   conjunction is not used to suggest what conjunction to try next.

The Resolution procedure [Robinson 65] avoids these inefficiencies by
not going straight to variable free clauses. Instead the search for a
contradiction is carried out on the original clauses with substitutions being
made when they will allow the continuation of the search. Thus

– Contradiction tests are carried out on several variable free
   conjunctions, in parallel.

– Conjunctions which could not play a part in the contradiction are
   implicitly eliminated by making substitutions which exclude them.

The Resolution contradiction test consists of using the resolution rule of
inference to derive new clauses from the original ones, in an attempt to
derive the empty clause (which is equivalent to f, see section 5.2). A
conjunction which contains f is unsatisfiable, since it will be false whatever
assignments are made to the other clauses.

For this to imply the unsatisfiability of the original clauses we must show
that the resolution rule cannot derive a clause which would make a
satisfiable conjunction into an unsatisfiable one: that is, any new clause
must be a logical consequence of the old ones. This is the *Soundness
Theorem*, so called, because it shows that resolution is not a faulty rule.

In addition we will want to know that if the original clauses are unsatisfiable then the contradiction test will eventually succeed, i.e. that the empty clause will eventually be derived. This is the Completeness Theorem, so called, because it shows that resolution is all you need. We prove both these theorems in section 16.6 below.

16.6 The Soundness and Completeness of Resolution

The Resolution process of deriving new clauses from old will eventually derive the empty clause if and only if the original clauses were unsatisfiable. The 'only if' part of this is the *Soundness Theorem:* that we cannot produce an unsatisfiable conjunction of clauses, and in particular, one containing the empty clause, from a satisfiable one.

Theorem 6: Soundness
    If a formula, A, in conjunctive normal form, is satisfiable and if C is derived from two clauses in A by resolution then A & C is satisfiable.

Proof: Suppose A is satisfiable then it has a model, I.

    Let C be derived from the clauses

1. $C' \vee P'_1 \vee ... \vee P'_m$ and
2. $C'' \vee \sim P''_1 \vee ... \vee P''_n$
such that $C = (C' \vee C'')\varphi$ where φ is the most general unifier of the P's and P''s. 1. and 2. are true in I, so all instances of them are true in I. true in I.
Let Θ be a typical substitution making $(C' \vee C'')\varphi\Theta$ variable free. There may be variables in the P's and P''s which are not in C' or C'', but $\varphi\Theta$ can be extended to a substitution $\varphi\Theta\psi$ which makes both 1. and 2. variable free. These instances are true in I, since I is a model of A.

From the semantic tree for ∨ applied to 1. the following two cases arise.

1. $C'\varphi\Theta\psi$ is true in I, in which case $(C' \vee C'')\varphi\Theta$ is true in I.

2. Some $P'_i \varphi\Theta\psi$ is true in I. But $P'_j \varphi = P''_k \varphi$ for all j and k, so all $P'_j\varphi\Theta\psi$ are true and all $\sim P''_k \varphi\Theta\psi$ are false in I. Hence $C''\varphi\Theta\psi$ must be true in I, so $(C' \vee C'')\varphi\Theta$ is true in I.

Either way $(C' \vee C'')\varphi\Theta$ is true in I so $(C' \vee C'')\varphi$ is true in I and I is a model of C.   QED

Exercise 53: f is false or, equivalently, the empty clause. unsats(S) means that S is an unsatisfiable set of clauses. El ∈ S means that El is a member of set S. resolvants(N,S) is the set of clauses that can be derived from S in 0 to N resolutions.

(a) Express the following axioms in clausal form (Kowalski form).

$\forall$S $\forall$N [unsats(resolvants(N,S)) $\rightarrow$ unsats(S)]

$\forall$S [f ϵ S $\rightarrow$ unsats(S)]

(b) Negate the following conjecture and express it in clausal form

$\forall$S $\forall$N [f ϵ resolvants(N,S) $\rightarrow$ unsats(S)]

(c) Using the goal clause of (b) as top clause draw a Lush Resolution search tree for the clauses from (a) and (b).

The 'if' part is the Completeness Theorem: that is, if we start with an unsatisfiable set and go on deriving new clauses by resolution, we will eventually derive the empty clause.

The idea of the proof is that each time we conjoin a newly derived batch of clauses to the unsatisfiable conjunction its failure tree gets smaller. Eventually its failure tree will be the trivial tree consisting of just the root node. Since only the empty clause, f, can have the trivial failure tree, f must be one of the conjuncts, i.e. at some stage f was derived.

Why does the failure tree keep getting smaller? Because we can always find two failure nodes (i.e. tips of the tree) with the same parent node. If we take the clauses which fail at these nodes and resolve them we get a clause which fails at the parent (or above) and so the failure tree can be pruned of these two tips (at least) to form a smaller failure tree for the new conjunct. The parent node, under these circumstances, is called an *inference node*.

We will want to make precise the idea of a 'newly derived batch of clauses'. These will be all possible resolvents of all clauses from A, the original formula. When these are conjoined to A the result is denoted $R(A)$. $R(R(R(...(A)...)))$ n times is denoted R^n (A). We will now prove:

Theorem 7: Completeness

     If A is an unsatisfiable formula in conjunctive normal form then there is an n such that f is a clause of R (A)

Proof: Suppose A is unsatisfiable then by lemma 4 it has a finite failure tree T.

Either T is the trivial tree, in which case f is a clause of A and n=0.

Or T is not the trivial tree. Pick a failure node (i.e. a tip), N' of maximum depth in T. Let N be the parent of N' (It has one, since T is not trivial). Since T is minimal, N is not a failure node. Let N" be its other daughter. Since N' is of maximum depth then N" must also be a failure node. Hence N is an inference node.

Without loss of generality let the arc from N to N assign f' to P and the arc from N to N" assign t to P (see figure 16-3).

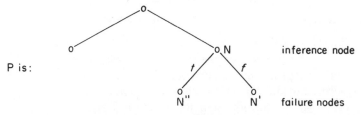

P is:

inference node

failure nodes

Figure 16-3: Part of Failure Tree for A

N' and N" are failure nodes of clauses from A, i.e. there are instances of two clauses from A made false by the interpretations defined by the branches terminating at N' and N". Since both these clauses are true at N (N is not a failure node) these instances must contain the literals P and ~P, respectively, i.e. they can be put in the form:

1. $(C' \vee P'_1 \vee ... \vee P'_m)\varphi'$ and
2. $(C'' \vee \sim P''_1 \vee ... \vee \sim P''_n)\varphi''$

where $C' \vee P'_1 \vee ... \vee P'_m$ and $C'' \vee \sim P''_1 \vee ... \vee \sim P''_n$ are clauses of A, $P'_i \varphi' \equiv P''_j \varphi'' \equiv P$ and P is not contained in $C'\varphi$ nor ~P in $C''\varphi$

Since the P's and P"s have a common instance, they must have a most general unifier φ and hence $(C' \vee C'')\varphi$ must be derivable by resolution from 1. and 2. i.e.

$(C' \vee C'')\varphi \, \epsilon \, R(A)$

Now $(C' \vee C'')\varphi'\varphi''$ is an instance of $(C' \vee C'')\varphi$ and all the literals in it are assigned f at some arc above N (because 1 and 2 fail at N' and N", respectively). Therefore, N is either a failure node for $(C' \vee C'')\varphi$ or its failure node is higher in the tree. In either case the failure tree for R(A) can be formed from T by pruning at least N' and N" and is therefore smaller than T.

Since A is unsatisfiable and is a subconjunct of R(A) then R(A) is unsatisfiable. The process can be repeated, with the failure tree for R^i (A) becoming smaller and smaller, until for some n the failure tree for R^n (A) is the trivial tree.   QED

Exercise 54: Using the notation of exercise 53, express the Soundness and Completeness theorems for sets of clauses.

16.7 Summary

In this chapter we have proved three theorems of Mathematical Logic: Herbrand's theorem and the Soundness and Completeness of Resolution.

These have suggested a procedure for showing that a conjecture, Thm, is a logical consequence of a set of axioms, Ax. The correctness of this procedure relies on the following chain of reasoning.

Thm is a logical consequence of Ax
      iff (consider rule for universal quantifier)
Thm' is a logical consequence of Ax',
where Thm' and Ax' are the closures of Thm and Ax respectively.
      iff (consider semantic trees for & and $\sim$)
Ax' & $\sim$Thm' is unsatisfiable
      iff (by theorem 6)
nf(Ax' & $\sim$Thm') is unsatisfiable
      iff (by lemma 2
nf(Ax' & $\sim$Thm') has no Herbrand Models
      iff (by theorems 6 and 7)
The empty clause can be derived from nf(Ax' & $\sim$Thm') using the resolution rule of inference.

Further Reading Suggestions

The description of several refinements of Resolution, and proofs of their soundness and completeness, can be found in [Chang and Lee 73, Loveland 78]. [Loveland 78] is a mathematically demanding book.

17

Pattern Matching

Most computational reasoning processes involve *pattern matching*, e.g. to recognise that an axiom is applicable to the current problem. In resolution this role is played by unification. Unification finds a substitution which will make two or more propositions identical. For instance, the recognition that the axiom

$$U=V \ \& \ W=V \to U=W$$

can be applied to (i.e. resolved with) the goal clause

$$b=a \to$$

consists mainly of finding the substitution

$$\{b/U, a/W\}$$

which will make $U=W$ and $b=a$ identical.

17.1 One Way Matching

Finding the above substitution is a particularly simple example of unification, called one way matching, because the substitution need only be applied to one of the expressions, $U=W$. We will call this expression the *pattern* and the one to which no substitutions are to be applied, the *target*. One way matching is all that is required for many mathematical reasoning processes, e.g. the application of rewrite rules (see chapter 9) and the check for subsumption. Note that, in subsumption, it is vital that the target be allowed to contain variables even though it is not possible to substitute for them.

It is easy to write a computer program to do one way matching. We will call it the *match* procedure.

(a)The first step of the match procedure will be to compare the two expressions and find a pair of subexpressions which disagree. For instance, if the pattern were X=X and the target 1+2=1 then <X, 1+2> is a suitable pair and so is <X, 1>. We will call such pairs *disagreement pairs*. They can be found by searching the expression trees of the two expressions, simultaneously, by say depth first search, until two nodes with different labels are reached. The disagreement pair is then the pair of subtrees dominated by these nodes (see figure 17-1).

(b)If the disagreement pair is <Var, term>, where Var is a variable, then the disagreement can be resolved with the substitution term/Var}. This substitution should be remembered as part of the eventual output of the procedure. It should also be applied to the pattern to prevent the variable Var subsequently being matched to some different term.

(c)The search for disagreement pairs should now continue. If a pair is found then the above process of forming a substitution should be repeated. If no pair is found, then the pattern and target must be identical; the procedure may be exited and the substitution found so far output as the result of the procedure.

(d)If a disagreement pair is ever found in which the first member is not a variable then the match cannot succeed and should be abandoned forthwith.

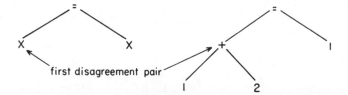

Figure 17-1: Finding Disagreement Pairs

Suppose that the first disagreement pair found in our example were <X, 1+2>. The substitution {1+2/X} would be formed and the pattern instantiated to 1+2=1+2. Continuing the search for another disagreement pair would uncover <1+2, 1>. Since 1+2 is not a variable the matching process would now fail.

The match procedure can be summarized by the following instructions

To match pattern and target given substitution, φ

1. If pattern and target are identical then succeed with output φ
2. Otherwise, let <t1,t2> be the first disagreement pair.
3. If t1 is a variable then call match on pattern {t2/t1} and target given φ {t2/t1}.
4. Else fail.

When it is first called *match* is, of course, given the empty set as the substitution φ

17.2 Combining Substitutions

The match procedure above calls for the substitution φ to be combined with the substitution {t2/t1}. In this section we describe how this combination is to effected. We will want the method of combining substitutions to have the property that applying two substitutions to an expression in succession produces the same result as applying the combination to the expression, i.e.

$$(\text{Expr}\,\varphi\,)\,\Theta \equiv \text{Expr}(\varphi\,\Theta) \qquad\qquad (i)$$

Unfortunately, the combination method which has this effect is a little messy.

The obvious first approximation is to take the union of the two sets which constitute the substitutions. And this is the correct thing to do when there is no overlap between the substitutions, as with {a/X, g(a)/Y} and {a/Z}, since

$$f(X,Y,Z) \{a/X, g(a)/Y\}) \{a/Z\}$$
$$\equiv f(a,g(a),Z) \{a/Z\}$$
$$\equiv f(a,g(a),a)$$

and

$$f(X,Y,Z) (\{a/X, g(a)/Y\} \cup \{a/Z\})$$
$$\equiv f(X,Y,Z) \{a/X, g(a)/Y, a/Z\}$$
$$\equiv f(a,g(a),a)$$

However, the substitutions may overlap in two ways: the second substitution may contain a pair t1/X, where X appears in the first substitution, either as a variable being replaced or in a term being substituted, e.g.

1. combine {a/X} and {b/X} or
2. combine {g(Y)/X} and {b/Y}

Consider case 2. first. Note that

f(X) {g(Y)/X}) {b/Y}
    $\equiv$ f(g(Y)) {b/Y}
    $\equiv$ f(g(b))

  whereas
f(X) ({g(Y)/X} $\cup$ {b/Y})
    $\equiv$ f(X) {g(Y)/X, b/Y}
    $\equiv$ f(g(Y))

and f(g(b)) is not identical to f(g(Y))

The answer is to apply the second substitution to the terms of the first before taking the union of the substitutions, e.g.

f(X) ({g(Y)/X} {b/Y})
    $\equiv$ f(X) ({g(Y){b/Y}/X} $\cup$ {b/Y})
    $\equiv$ f(X) {g(b)/X, b/Y}
    $\equiv$ f(g(b))

Now consider case 1. Note that taking the union of {a/X} and {b/X} will produce an ambiguous substitution {a/X, b/X}. Since the first substitution is applied first, the pair b/X will not get a look in and can be ignored, i.e.

f(X) {a/X}) {b/X}
    $\equiv$ f(a) {b/X}
    $\equiv$ f(a)

Hence {a/X}{b/X} should be {a/X}

The only way that {b/X} might get a look-in would be if the first substitution reintroduced X, but this is an example of case 2. and is already handled correctly, e.g.

f(X) {g(X)/X}) {b/X}
    $\equiv$ f(g(X)) {b/X}
    $\equiv$ f(g(b))
  and

f(X) ({g(X)/X} {b/X})
    $\equiv$ f(X) {g(X){b/X} / X}
    $\equiv$ f(X) {g(b)/X}
    $\equiv$ f(g(b))

We will call the method of combining substitutions outlined above combine.

The application of combine to two substitutions will be denoted, as above, by juxtaposing them, i.e. by writing them one after the other. The full definition of combine is given by the following procedure.

To combine substitutions Θ and φ

1. Replace each pair s/X in Θ by sφ/X to form Θ'.
2. Delete from φ each pair t1/Y,
          such that Θ' contains a pair s/Y, to form φ'.
3. Output the union of Θ' and φ'.

combine has the property (i) and is also associative, but we omit the proofs of these facts.

Exercise 55: Apply the combine procedure to the substitutions

(a)$\{1+2/X, U+2/Y\}$ and $\{V/Z, 3+2/W\}$

(b)$\{1+2/X, U+2/Y\}$ and $\{V/Y, 3+2/W\}$

(c)$\{1+2/X, U+Y/Y\}$ and $\{V/Y, 3+2/W\}$

Exercise 56: Apply one way matching to the following pairs of expressions. Determine whether the process fails or succeeds in each case. In the case of success determine the resulting substitution.

pattern	target
a) X=X	2=2
b) X=X	2+2=4
c) p(f(X,Y), Y)	p(f(a,g(b)), g(b))
d) X=b	a=Y

17.3 Unification

In the last example in exercise 56 we perceive the need for a two way matching process that could produce a substitution for application to both expressions, e.g. $\{a/X, b/Y\}$ whether applied to X=b or a=Y would produce the common instance, a=b. With only one way matching the two propositions are unmatchable. A suitable two way process is provided by *unification*.

At first sight it appears simple to amend the one way match procedure to do two way matching. We need only add the additional line:

3'. If t2 is a variable then call
          match on pattern and target $\{t1/t2\}$ given $\varphi\{t1/t2\}$.

and substitute 'unify' for 'match' throughout the procedure. But it is not quite as simple as that.

17.3.1 Symmetric Application of Substitutions

Firstly, despite having standardized variables apart initially, the substitution process may pollute the pattern with variables from the target and vice versa. Consider, for instance,

unify X=a and Y=Y given {}.

The first disagreement pair produces the substitution {Y/X} and calls

unify Y=a and Y=Y given {Y/X}.

Variable Y is now in both expressions. The second disagreement pair produces the substitution, {a/Y}. If this is only applied to the target, as in line 3' above then

unify Y=a and a=a given {a/X, a/Y}.

will be called and a redundant round of matching will be needed to complete the job.

Hence newly discovered substitution pairs must be applied to both expressions and lines 3 and 3' must be amended to:

3. If t1 is a variable then call
   unify exp1{t2/t1} and exp2{t2/t1} given φ{t2/t1}
3a. If t2 is a variable then call
   unify exp1{t1/t2} and exp2{t1/t2} given φ{t1/t2}

We have also replaced 'pattern' and 'target' with the more symmetric 'exp1' and 'exp2'.

17.3.2 Occurs Check

The second wrinkle is more subtle. Now that the same variable can occur in both expressions there is a danger of creating a looping substitution, in which a variable is bound to a term it occurs in, e.g. {Y+1/Y}. This will happen, for instance, if the current procedure is applied to

X=X and Y=Y+1 given {}.

Clearly these expressions have no common instance, but as defined above unify will not fail on them. It will loop forever, i.e.

unify $Y=Y$ and $Y=Y+1$ given $\{Y/X\}$

unify $Y+1=Y+1$ and $Y+1=(Y+1)+1$ given $\{Y+1/X, Y+1/Y\}$

unify $(Y+1)+1=(Y+1)+1$ and $(Y+1)+1=((Y+1)+1)+1$
    given $\{(Y+1)+1/X, (Y+1)+1/Y\}$

....

In such cases we would like unify to terminate with failure. It can be amended to do so by inserting an *occurs check* in lines 3 and 3', i.e.

3. If t1 is a variable and t1 does not occur in t2 then call unify on expl$\{t2/t1\}$ and exp2$\{t2/t1\}$ given $\varphi\{t2/t1\}$

The complete unification procedure is now:

To unify exp1 and exp2 given φ

1. If exp1 and exp2 are identical then succeed and output φ
2. Otherwise, let $<t1, t2>$ be the first disagreement pair.
3. If t1 is a variable and t1 does not occur in t2 then call unify on expl$\{t2/t1\}$ and exp2$\{t2/t1\}$ given $\varphi\{t2/t1\}$.
3'. If t2 is a variable and t2 does not occur in t1 then call unify on expl$\{t1/t2\}$ and exp2$\{t1/t2\}$ given $\varphi\{t1/t2\}$.
4. Else fail.

Exercise 57: Apply unify to the following pairs of expressions. Determine whether the procedure fails or succeeds in each case. In the case of success determine the resulting substitution.

exp1	exp2
a) $X=b$	$a=Y$
b) $X=b$	$Y=a$
c) $p(X,a)$	$p(f(Y),Y)$
d) $p(X,g(X))$	$p(f(Y),Y)$
e) $(a+X)+b$	$a+Y$

17.3.3 General Unification

The unify procedure unifies two propositions. Full resolution and the factoring rule both require the simultaneous unification of a *set* of propositions. Fortunately, pairwise unification can be easily extended to unification of a set of expressions. The following modifications are required.

- exp1 and exp2 must be replaced with a set of expressions, expset.

- As substitutions are applied to this set it will be reduced in size. If it ever becomes a singleton the process terminates with success.

- The disagreement pair, <t1,t2>, must be replaced by a disagreement set, disagset, of corresponding subexpressions from each member of expset.

- The procedure can continue, with a new substitution pair, at step 3 if one of the elements of disagset is a variable and another is a term not containing that variable.

Thus the general unification procedure is:

To gen-unify expset given φ

1. If expset is a singleton then succeed and output φ
2. Otherwise, let disagset be the first disagreement set of expset.
3. If disagset contains a variable V and a term t1 and V does not occur in t1 then call gen-unify on expset$\{t1/V\}$ given $\varphi\{t1/V\}$

4. Else fail.

17.3.4 Theoretical Properties of gen-unify

The resolution rule puts the following demands on the unification procedure.

- Unification must succeed in unifying expset precisely when there is a substitution φ such that expsetφ is a singleton, i.e. precisely when expset is *unifiable*.

- Unification should return the most general unifier , φ of expset. That is, there must be no Θ such that expsetΘ is a singleton and expset is an instance of expsetΘ, unless expsetΘ is also an instance of expsetφ i.e. unless φ and Θ are alphabetic variants.

We are happy to report, without proof, that gen-unify has both these desirable properties. A proof can be found in [Chang and Lee 73] pp79-80.

Another nice property of unification, is that up to alphabetic variance, a unifiable set of expressions has only one most general unifier. So we can talk about *the* most general unifier.

17.4 Building-In Axioms

You may have been surprised at which of the pairs of expressions in exercise

57 did not unify. Did you think that <X=b, Y=a> and <(a+X)+b, a+Y> would unify? Why was it that the unify procedure did not suggest the substitutions {b/Y, a/X} and {X+b/ Y}, respectively?

The answer is that in each of these cases special properties of the symbols involved must be exploited if they are to match, namely the symmetry of = and the associativity of +. After all you would not expect the neutral versions of these examples: <p(X,b), p(Y,a)> and <f(f(a,X),b), f(a,Y)>: to match! unify has no access to special properties of predicates and functions. This section is about how it can be given such access.

But why would we want to modify the unify procedure? True, clauses like

$$(a+X)+b>c \rightarrow \text{ and } \rightarrow a+Y>c$$

cannot be resolved directly, but the first can be transformed by an application of the associativity axiom and its descendant resolved with the second (see figure 17-2).

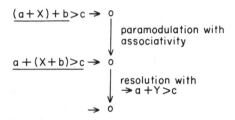

Figure 17-2: Indirect Resolution of Non-Unifiable Propositions

However, just as with the indirect proofs using the equality axioms that we encountered in section 5.3, this sort of longwindedness detracts from the naturalness of the proofs as well as making them longer. We will see how to cure this by giving the unification procedure access to special properties of symbols, like the associativity of +.

17.4.1 Associative Unification

As an example we will consider how to build the associativity of + into the binary unification procedure, unify.

What do we mean by *building-in*? How will we know when we have succeeded? We can adopt the same criterion here as we did with paramodulation, that is: we want to modify the unification procedure in such a way that the associativity axiom for + can be deleted and yet precisely those formulae which were provable before will be provable still. We will call the modified procedure, *assoc-unify.*

How can we start? It would be nice if the procedure for finding disagreement pairs ignored disagreements like $<(a+X)+b, a+Y>$ and went on to find $<X+b, Y>$ instead. This effect can be achieved by keeping all expressions in an *associative normal form*, e.g. right bracketed. We will denote the right bracketed normal form of an expression, exp1, by rbnf(exp1). So $(a+X)+b$ should be normalized to $a+(X+b)$. If this is done then the regular disagreement finding procedure can be retained.

Substitutions tend to de-normalize expressions, e.g. although $a+(X+b)$ is in normal form, applying the substitution $\{c+d/X\}$ to it transforms it to $a+((c+d)+b)$, which is not in normal form. So expressions must always be re-normalized after instantiation.

For comparing variable free expressions it suffices to put them in normal form and unify them, but for expressions containing variables further wrinkles are required. This is because of the possibility of expressions with variables being de-normalized. For instance, $X+Z$ and $a+(Y+b)$ are both in normal form. The regular unification procedure would output $\{a/X, Y+b/Z\}$, but this is not sufficient. We will show that the substitutions $\{a+Y/X, b/Z\}$ and $\{a+U/X, U+V/Y, V+b/Z\}$ must also be output, if the associativity axiom is to be made totally redundant.

Consider these terms in the context

$\rightarrow q(X+Z, Z)$ and
$q(a+(Y+b), b) \rightarrow$

These clauses are not regularly unifiable, but the empty clause can be deduced from them with the aid of associativity, resolution and paramodulation (see figure 17-3).

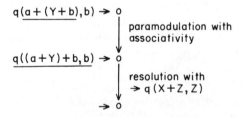

Figure 17-3: The Need for Non-Standard Unifiers

If associativity is to be deleted and yet the empty clause still be deducible from the rump, we will want assoc-unify to unify $X+Z$ and $a+(Y+b)$ with unifier $\{a+Y/X, b/Z\}$.

Now consider the same terms in a different context.

$\rightarrow$ r(X+Z, X, Z) and
  r(a+(Y+b), a+U, V+b) $\rightarrow$

Again the clauses are not regularly unifiable, but the empty clause is deducible from them with the aid of associativity, paramodulation and resolution.

Exercise 58: Check this.

So we will also want assoc-unify to unify X+Z and a+(Y+b) with unifier {a+U/X, U+V/Y, V+b/Z}.

We can already see that assoc-unify should not produce a single most general unifier, but several unifiers, none of which is a special case of the others. In fact we will see that an infinite number of most general unifiers is required in general.

The unification procedure can be modified to produce multiple unifiers by allowing it a choice when resolving disagreements. If <Var, term> is a disagreement pair we must allow either the substitution {term/Var} or the substitution {term+U/Var} to resolve the disagreement, where U is a new variable. By a judicious choice of substitution during unification of X+Z and a+(Y+b) the modified procedure can be made to produce each of the three desired unifiers.

Consider, for instance, how the third unifier might be produced from a call of

assoc-unify X+Z and a+(Y+b) given {}

If left first/ depth first search is used to find a disagreement pair it will return <X,a>. X is a variable and a is not so we have two ways of resolving the disagreement: to form substitution {a/X} or {a+U/X}. If we choose the latter possibility we generate the call

assoc-unify a+(U+Z) and a+(Y+b) given {a+U/X}

(a+U)+Z having been normalized. The disagreement pair of these expressions is <U,Y>. Since both of these are variables there are three ways of resolving the disagreement: to form substitution {Y/U}, {Y+W/U} or {U+V/Y}. The fourth possibility, {U/Y} is redundant. Again we will choose the last possibility and generate the call

assoc-unify a+(U+Z) and a+(U+(V+b)) given {a+U/X, U+V/Y}

after normalizing a+((U+V)+b). The disagreement pair of these expressions is <Z, V+b>. Z is a variable and V+b is not so we have two ways of resolving the disagreement: to form substitution $\{V+b/\ Z\}$ or $\{(V+b)+Z'/Z\}$. This time we will choose the first possibility and generate the call

assoc-unify a+(U+(V+b)) and a+(U+(V+b)) given
$\{a+U/X, U+V/Y, V+b/Z\}$

which will succeed, since the two expressions are identical, and output the unifier

$\{a+U/X, U+V/Y, V+b/Z\}$
as desired.

Exercise 59: Show that by making different choices during this process the other two substitutions can also be generated.

So the associative unification procedure can be summarized as:

To assoc-unify exp1 and exp2 given ϕ
1. If exp1 and exp2 are identical then succeed and output ϕ
2. Otherwise, let t1 and t2 be the disagreement pair.
3". If t1 and t2 are both variables then let $\ominus$ be either
   $\{t2/t1\}$ or $\{t2+V/t1\}$ or $\{t1+V/t2\}$ and call
   assoc-unify on rbnf(exp1$\ominus$) and rbnf(exp2$\ominus$) given ϕ $\ominus$.
3. If t1 is a variable and t2 is not and if t1 does not occur in t2
   then let $\ominus$ be either $\{t2/t1\}$ or $\{t2+V/t1\}$ and call
   assoc-unify on rbnf(exp1$\ominus$) and rbnf(exp2$\ominus$) given ϕ $\ominus$.
3'. If t2 is a variable and t1 is not and if t2 does not occur in t1 then let $\ominus$
   be either $\{t1/t2\}$ or $\{t1+V/t2\}$ and call assoc-unify on
   rbnf(exp1$\ominus$) and rbnf(exp2$\ominus$) given $\phi\ominus$.
4. Else fail.

where V is a newly created variable.

17.4.2 Theoretical Properties of assoc-unify

This procedure does the job it was intended to: that is, a set of clauses, S, containing the associativity axiom, is unsatisfiable iff the empty clause can be derived from S minus the associativity axiom by resolution using assoc-unify. A proof of this result can be found in [Plotkin 72].

We have already seen that assoc-unify is capable of outputting more than one most general unifier. In consequence, propositions are assoc-unifiable in more than one way and this adds to the number of ways in which two clauses can be resolved.

There are occasions on which two propositions may have an infinite number of unifiers and thus two clauses may have an infinite number of resolvants. This will cause breadth first search to be an incomplete search strategy, since there will be an infinite number of nodes all at the same depth in the search tree. In such circumstances it is best to regard unification as a part of the search process. For instance, we may incorporate unification into a depth first search by exercising one of the choices at steps 3, 3' or 3", and then being prepared to back up later and remake it.

Here are two expressions with an infinite number of unifiers: $g(X,X+a)$ and $g(Y,a+Y)$. If we denote $a+(a+....+a)...))$ (n times) by n.a then the substitutions $\{n.a/X, n.a/Y\}$ for $n=1, 2, 3, ...$ are all most general unifiers.

We can see this should be so by noting that the empty clause can be derived from

$\rightarrow p(g(X,X+a), X)$ and
$p(g(Y,a+Y), n.a) \rightarrow$

using associativity, paramodulation and resolution.

Exercise 60: Check this in the case $n=2$

The following sequence shows how the assoc-unify procedure generates a variant of the $n+1$st of these unifiers.

assoc-unify $g(X,X+a)$ and $g(Y,a+Y)$ given $\{\}$

assoc-unify $g(Y,Y+a)$ and $g(Y,a+Y)$ given $\{Y/X\}$

assoc-unify $g(a+Z1,a+Z1+a)$ and $g(a+Z1,a+a+Z1)$ given $\{a+Z1/X, a+Z1/Y\}^*$

....
assoc-unify $g(n.a+Zn,n.a+Zn+a)$ and $g(n.a+Zn,(n+1).a+Zn)$ given $\{n.a+Zn/X, n.a+Zn/Y\}$

assoc-unify $g((n+1).a ,(n+2).a)$ and $g((n+1).a, (n+2).a)$ given $\{(n+1).a/X, (n+1).a/Y\}$

assoc-unify is not a decision procedure for associative unification, because it may not terminate. We have seen that it may continue for ever turning out most general unifiers of associatively unifiable expressions. There are also expressions which are not associatively unifiable, but where the assoc-unify procedure never stops, e.g. $g(X,X+a)$ and $g(Y,b+Y)$.

However, the question of whether two expressions are associatively

---

*Since all expressions are in normal form the brackets are redundant and have been omitted.

unifiable is decidable, that is there is a procedure, *assoc-unifiable*, which given two expressions does terminate and outputs 'yes' or 'no' according to whether or not they are associatively unifiable. This procedure could be coupled to assoc-unify, so that assoc-unify was only called in the case that assoc-unifiable returned 'yes'. But even this combined procedure would not terminate when there were an infinite collection of most general unifiers to be found.

This distinction between knowing when two expressions are unifiable and knowing what their unifiers actually are is an important one. We will meet it again in lambda calculus unification.

Unification procedures are known which build-in the axioms of commutativity, idempotency, distributivity, homomorphism, and several combinations of these axioms. A survey of the theoretical results known about these procedures can be found in [Raulefs et al 78].

17.5 Lambda Calculus Unification

At the beginning of chapter 5 we said that the resolution and paramodulation rules of inference were only applicable to First Order Theories. The only obstacle to dealing with higher order theories is in designing a unification procedure which can treat variable functions, functionals etc correctly.

17.5.1 F-Matching

The regular unification procedure, unify, can be easily adapted to higher order expressions, e.g. so that it will unify F(0) and sin(X) producing most general unifier {sin/F, 0/X}. We will need to adjust the procedure for finding disagreement pairs so that if the root node of one subexpression is a variable then only the two root nodes are returned (see figure 17-4).

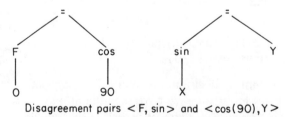

Disagreement pairs < F, sin > and < cos(90), Y >

Figure 17-4: Modified Disagreement Pair Finding

This adapted procedure is sometimes called F-Matching.

17.5.2 Building-in the Laws of Lambda Calculus

F-Matching is not generally considered adequate for higher order theorem proving because it cannot match expressions like

$$F(Y) \tag{ii}$$
$$\text{and } \sin(X).e^X \tag{iii}$$

where the function that F is to be matched to is not explicitly named, but must be formed by lambda abstraction . There are two different substitutions, which applied to F(Y) and $\sin(X).e^X$, will produce equal expressions. These are

$$\{ \lambda Z \ \sin(Z).e^Z \ /F, \ X \ /Y \}$$
$$\text{and}$$
$$\{ \lambda Z \ Z \ /F, \ \sin(X).e \ /Y \}$$

The first substitution is an example of *imitation*, so called because F is instantiated in order to behave like (or imitate) a lambda abstracted version of (iii). The second substitution is an example of *projection*, so called because F is defined as an identity (or projection) function whose value is its parameter, i.e. F(Z)=Z. The matching problem is shifted to this parameter.

To see that these substitutions do the job consider the result of applying them to F(Y). The first produces

$$(\lambda Z \ \sin(Z).e^Z)(X) \tag{iv}$$

and the second

$$(\lambda Z \ Z)(\sin(X).e^X) \tag{v}$$

both of which rewrite to

$$\sin(X).e^X \tag{vi}$$

17.5.3 The Laws of Lambda Calculus

In asserting that (iv) and (v) rewrite to (vi) we have appealed to the intuitive meaning of lambda abstraction. These rewritings cannot be *proved* correct by the laws of logic introduced so far: we first need to introduce a new rule of inference, called the *Beta Rule*, which allows us to substitute the parameter of a lambda abstraction for its bound variable and drop the lambda. We embed the lambda abstraction in some arbitrary expression, a.

Beta Rule:

$$\frac{a[(\lambda X \ e(X))(t)]}{a[e(t)]}$$

provided no confusion of bound variables is created by the replacement of X by t in e(X).

Similarly, in order to see that

$$\lambda X \ \sin(X).e^X \ \text{and} \ \lambda Y \ \sin(Y).e^Y$$

are essentially equal we will need the *Alpha Rule* of inference.

Alpha Rule:

$$\frac{a[\lambda X \ e(X)]}{a[\lambda Y \ e(Y)]}$$

provided that: Y is not already a free variable of e(X);
and X and Y have the same type;
and no confusion of bound variables occurs.

Lastly, in order to see that

$$\lambda X \sin(X) \ \text{and} \ \sin$$

are essentially equal, we must introduce the *Eta Rule*

$$\lambda X \ e(X) = e$$

provided e does not contain X, since such an X would get caught up in the scope of the λ.

17.5.4 The Lambda Unifiability Procedure

We will now give a unifiability procedure, *lambda-unifiable,* for Typed Lambda Calculus, due to Gerard Huet [Huet 74], which builds in the Alpha, Beta and Eta Rules. Since these laws of Lambda Calculus are built into the unification procedure it can be used, with resolution, to form a complete theorem proving system for Typed Lambda Calculus, whereas if F-Matching alone were used then the laws of Lambda Calculus would need to be used to supplement resolution.

Just as in assoc-unify, the expressions to be unified must be kept in a normal form. The normal form required is that all applications must be eliminated, i.e. no expression must contain a subexpression of the form $(\lambda X \ e)(t)$. Such subexpressions can always be eliminated with the Beta Rule, with the Alpha Rule being used, where necessary, to make the Beta Rule legal. lambda-unifiable preserves this normal form, so it is only necessary to normalize expressions once, before unification. An expression in normal form can always be written in the form

$$\lambda X_1 \ldots \lambda X_n f(e_1, \ldots, e_p) \hspace{4cm} \text{(vii)}$$

where f is a variable or constant, called the *head*.

lambda-unifiable will work on a set of pairs of expressions: trying to show each pair unifiable. We will call this set, *expr-pairs*. With previous unification procedures a pair of expressions were unifiable when one of them was a variable. In lambda-unifiable we will need a slightly different notion. A pair of expressions are unifiable when one of them is *flexible*, i.e. when its *head* is a variable which is different from all the X_i, i $\in \{1,..,n\}$. An expression which is not flexible, i.e. whose head is either constant or one of the X_i, is called *rigid*.

Rigid/flexible pairs will either fail or be unified with either imitation or projection. Rigid/rigid pairs will either fail or their parameters will be paired off and added to expr-pairs. Flexible/flexible pairs can always be matched and in lots of ways. To avoid a potentially infinite explosion at this point we will merely record that the match succeeds without saying exactly which substitution will unify the expressions. Thus lambda-unifiable tests for unifiability rather than producing substitutions.

To test whether the pairs in expr-pairs are lambda-unifiable

1. Check that the expressions in each rigid/rigid pair in expr-pairs have the same head,
   (modulo changes of bound variable), i.e.
   For each rigid/rigid pair, $<e, e'>$, where

   $$e = \lambda X_1 \ldots \lambda X_n f(e_1, \ldots, e_p)$$
   $$e' = \lambda X_1' \ldots \lambda X_{n'}' f'(e_1', \ldots, e_{p'}')$$

   If $n \neq n'$ then fail
   If $f \neq \lambda X_1' .. \lambda X_{n'}' f'(X_1, \ldots, X_n)$ then fail
   Otherwise $p = p'$. Replace $<e, e'>$ in expr-pairs with
   $\quad <e_i, e_i'>$ for all i $\in \{1, \ldots, p\}$
   where $e_i = \lambda X_1 \ldots \lambda X_n e_i$
   $\quad e_1' = \lambda X_1' \ldots \lambda X_n' e_i'$

2. If all pairs in expr-pairs are flexible/flexible then succeed.

3. Reverse all flexible/rigid pairs into rigid/flexible pairs.

4. Find substitutions for all rigid/flexible pairs, $<e, e'>$, before continuing.
   Let
   $$e = \lambda X_1 \ldots \lambda X_n f(e_1, \ldots, e_p) \text{ and}$$

$$e' = \lambda X_1' \ldots \lambda X'_n \, {}'F'(e_1', \ldots, e_p')$$
where F' is a variable not equal to any X_i'.

There are two ways to do this: by imitation or projection.

4a. Imitation:
   Let φ be the substitution:
   $\{ \lambda Y_1 \ldots \lambda Y_{p'} \, f(g_1, \ldots, g_{p-p'}) \, /F' \}$
   where $g_i = G_i(Y_1, \ldots, Y_{p'})$
   and each G_i is a new variable.

4b. Projection:
   Let φ be the substitution:
   $\lambda Y_1 \ldots \lambda Y_{p'} \, Y_i(g_1, \ldots, g_m) \, /F' \}$
   where $g_i = G_i(Y_1, \ldots, Y_{p'})$
   and each G_i is a new variable
provided the type of Y_i is appropriate.

In each case apply φ to each expression in expr-pairs and go back to step 1.

To decide whether a pair of expressions, expr and expr', are unifiable, lambda-unifiable is applied to the singleton, $\{<\text{expr}, \text{expr}'>\}$.

17.5.5 Theoretical Properties of lambda-unifiable

As noted above, lambda-unifiable builds in the laws of the Lambda Calculus: the Alpha, Beta and Eta Rules. A pair of expressions are unifiable, modulo these laws, iff lambda-unifiable would succeed on them.

Note that lambda-unifiable searches a tree of ways of unifying expressions. The choice point comes in step 4, when either imitation or projection can be used. This reflects the fact that a pair of expressions may be lambda unifiable in several different ways, yielding a different substitution each time. Some pairs of expressions have an infinite number of lambda unifiers. This infinite collection of lambda unifiers may be redundant: that is it may contain some unifiers which are not most general. It is known that there is no way to design a lambda unification procedure which produces non-redundant sets of unifiers. In contrast lambda-unifiable is non-redundant, and this is only possible because it eschews the search for sets of unifiers and just tests for unifiability.

lambda-unifiable is a semi-decision procedure : that is if a pair of expressions is unifiable then lambda-unifiable will terminate, but if they are not unifiable then lambda-unifiable may continue for ever. It is not possible to do better than this: the problem is basically a semi-decidable one. This is similar to the situation for Resolution theorem provers.

17.6 Summary

Pattern matching is a vital part of most computational reasoning processes, including Resolution theorem proving. There are a wide variety of pattern matching procedures: one way matchers, two way matchers (unifiers) and matchers which build in various axioms. These latter include matchers which build in associativity and matchers which build in the laws of the Typed Lambda Calculus. Better behaved matchers can sometimes be obtained by only testing for unifiability rather than outputting the unifying substitutions.

Further Reading Suggestions

[Raulefs et al 78] is a survey of unification procedures with an extensive bibliography.

18

Applications of Artificial Mathematicians

In this book we have discussed how to build computational models of various aspects of mathematical reasoning. We now consider whether these models have any practical significance: whether we can find *applications* for our artificial mathematicians.

The applications we will consider fall under three headings.

- ☐ *Technological:* the provision of aids for people who make professional use of mathematical reasoning, e.g. mathematicians themselves, scientists and engineers. We consider these applications in sections 18.1 and 18.2.
- ☐ *Educational:* the provision of models to help in teaching Mathematics. This includes expert models as part of a tuition system and faulty models to help the teacher diagnose student errors. We consider these applications in sections 18.3 and 18.4.
- ☐ *Scientific:* the provision of tools for the scientist interested in modelling or studying reasoning. This includes workers in Artificial Intelligence who are modelling inference in some other area of cognitive activity and psychologists interested in explaining human problem solving behaviour. We consider these applications in sections 18.4 and 18.5.

The rest of this chapter gives examples of each of these kinds of applications, starting with the technological ones.

18.1 Algebraic Manipulation Systems

The classic application of Mathematics is to Physics and Engineering. Physical systems are modelled with algebraic expressions: equations and inequalities; differential and non-differential; numeric and matrix. Some of the algebraic expressions which need to be solved are very large. They may need several pages of paper just to write down. Manipulating such expressions by hand, not only takes a long time, but usually introduces errors into the computation making the result worthless.

Computers have come to the rescue by offering a range of *Algebraic Manipulation Programs*. The best known of these programs are MACSYMA and REDUCE. Below we outline the MACSYMA program.

MACSYMA grew out of the work on theorem proving. Its predecessors were two programs to do symbolic integration called SAINT [Slagle 63] and SIN [Moses 67]. SAINT used the techniques of heuristic search. The root of the search tree was the integral to be solved. Arcs were built from this for each applicable integration method, leading to nodes for the sub-problems each of these methods introduced. SIN was developed to overcome the inefficiences of SAINT. It had more of the flavour of the Boyer/Moore program or of PRESS, in that it first conducted an analysis of the integration problem before bringing to bear an appropriate method. Finally the MACSYMA program grew from the addition to SIN of a variety of other algebraic manipulation abilities.

The current abilities of MACSYMA include:

- the symbolic integration and differentiation of expressions;

- the reduction of expressions to various normal forms;

- the solving of equations;

- the manipulation of matrices and

- the summing of infinite series.

A typical session with MACSYMA is displayed in figure 18-1.

C1 (X**2−Y**2) * (Z**2+2*Z) / ((X+Y)*W)@

D1 $\dfrac{(X^2-Y^2)(Z^2+2Z)}{W(Y+X)}$

C2 RATSIMP(D1)@

D2 $\dfrac{(X-Y)Z^2 + (2X-2Y)Z}{W}$

C3

Figure 18-1: A Typical MACSYMA Session

Lines beginning with a C were typed by the user and those beginning with a D by MACSYMA. The double asterisk indicates exponentiation and the @ sign is a sort of 'Roger and out'. When an algebraic expression is typed in (e.g. line C1), it is simplified slightly, echoed in pretty printed format and given a label (e.g. D1). This label can then be used to refer to the expression, as it is in line C2. Here a request is made to put the expression in a canonical form called, Rational Function Form, and abbreviated as RATSIMP. This canonical form is then labelled D2 and may be further manipulated.

The mathematical reasoning techniques embedded in MACSYMA include the use of sets of rewrite rules to define user specific normal forms and the use of pattern matching procedures. Although search techniques played a large role in MACSYMA's predecessors, they have now been expunged from MACSYMA itself. Its current integration method is based on a decision procedure due to Risch and Norman.

18.2 Automatic Theorem Proving

Despite the criticisms levelled at uniform proof procedures in chapter 7 they can still be used to prove interesting theorems in certain areas of Mathematics. This has been demonstrated in recent years by the work of Larry Wos, Ross Overbeek and Ewing Lusk [Wos 82]. They have developed a very efficiently coded Resolution theorem proving program, and have had mathematician colleagues use it to prove *open conjectures* in various branches of Mathematics.

A clue as to how this was possible can be gleaned by considering the areas from which the open conjectures were drawn. The Wos/Overbeek/Lusk theorem prover has established new theorems in: Ternary Boolean Algebra; Semigroups; Robbins Algebra; Equivantial Calculus; Finite Basis Problems; Knot Theory and Circuit Design. Note that you have never heard of some of these areas. They are mostly pretty new areas of Mathematics. Humans have not yet had centuries of involvement with them in which to develop methods, heuristics, intuitions, proof plans, etc. In such a case humans are reduced to the sort of exhaustive search, and faced with the sort of combinatorial explosions we discussed in chapter 7. Now the computer can compete on its own terms. When it comes to exhaustive search, computers have the bookkeeping skills, memory capacity and patience required, to dwarf human ability.

The Wos/Overbeek/Lusk theorem prover is one example of several computer programs designed as an aid for humans. The latest version of the Boyer/Moore theorem prover is another example. Most of the other systems are man/machine systems, i.e. they need and encourage the interaction of the human user in solving the problem, especially in making choices. The Wos/Overbeek/Lusk program is rare in that it provides minimal opportunity for human interaction and thus, apart from the initial setting of a few program parameters, its solutions are found unaided. It is also rare in that it solves open problems. Most of its rival programs have only been used, in an experimental mode, to reproduce solutions already known to the user, at least in outline.

The success of this program suggests that we might look forward to a range of powerful mathematician's aids; both man/machine and stand alone.

18.3 Computer Assisted Instruction

Another application of theorem-provers is to give automatic expert assistance to students who are learning to prove theorems. A good example of such an application is a program by Adele Goldberg [Goldberg 73] for teaching Group Theory.

The Goldberg tuition program work as follows.

1. The student is presented with a theorem of group theory and asked to prove it.

2. The student then types steps of the proof into the computer. A theorem proving component of the Goldberg program attempts to prove each step from the previous ones. If it cannot do this it complains to the student.

3. If the student asks for help the theorem prover is used to give hints.

4. If the student completes the proof the theorem prover is used to suggest alternative proofs.

There are several advantages to using a theorem prover to check the student's proof, at step 2, rather than, say, just comparing the student's proof with some prestored list of steps.

- The student is not restricted to a few 'approved' proofs, but may produce any correct one.

- The student need not give the proof in detail, but may omit trivial steps: the theorem prover will fill in the missing steps, provided the jumps are not too big.

The theorem prover is used to advantage at step 2, but it is indispensable at step 3. To give a hint relevant to the current state of the student's proof it is necessary to be able to perform inferences dynamically. The theorem prover finds all ways of extending the student's existing steps into a proof of the theorem. If this can be done in one step then hints are given on this step, otherwise a proof is chosen which makes maximum use of the student's partial proof and hints are given on the first step of this.

The Goldberg tuition program is an example of what is called Intelligent Computer Aided Instruction. It is based on the tenet that, a teacher must understand something if (s)he is to teach it successfully, even if the teacher is a computer program. This may seem obvious, but it is a tenet which is violated by the more conventional 'drill and practice' Computer Aided Instruction programs.

18.4 Understanding Students' Subtraction Errors

All the examples of applications considered so far have involved the modelling of expert performance. But we can also build models of novice performance and include in them the making of errors. If these errors explain how student errors can arise then they can also find educational application: by training teachers to diagnose a student's misconceptions.

The classic example of such work is the BUGGY program [Brown and Burton 78]. BUGGY is a tuition program for trainee teachers. It includes within it a program for doing subtraction. (And addition, but we will confine our attention to subtraction.) There is nothing very hard about writing a subtraction program: such programs have been standard since the earliest computers. But this is a subtraction program with a difference. The techniques of mathematical reasoning have been used to build a modular program. Bits of this program can be unplugged and other bits inserted to produce 'buggy' programs, i.e. programs which do some subtractions wrong. The Computer Science term for a fault in a program is a *bug*. Furthermore, there is large measure of agreement between those errors, which can be produced by simple modifications to the subtraction program, and those typically produced by real primary school children. BUGGY is based on the theory that a large number of student's subtraction errors, are not due to carelessness, but to carefully following a subtraction procedure containing bugs.

18.4.1 What BUGGY Does

The BUGGY program works as follows.

1. The teacher is given an example of an incorrect subtraction sum and invited to guess the underlying bug in the procedure which produced the sum.

2. (S)he can then either claim to have found the bug or experiment by proposing sums to BUGGY, which will run its buggy subtraction procedure on them.

3. When the teacher claims to know the bug (s)he is first invited to describe it in English and then to prove his/her knowledge by

predicting the BUGGY solution to five sums posed by the program.

4. If the teacher gets these sums 'right' then (s)he passes on to another buggy procedure or terminates the session. If (s)he gets any wrong then (s)he goes back to step 2.

A sample session is given in figure 18-2. The bits typed by the teacher are marked by a # sign at the start of the line. BUGGY is unable to analyse the English description of the teacher's hypothesis, and judges its correctness solely on the basis of the 5 test problems. Note that the teacher's first guess is wrong. This is usually the case and demonstrates the value of the experience of hypothesis making and testing, before the teacher is let loose on real students.

Here is an example of the bug.

```
   4 8
 - 1 9
 ─────
   3 9
 ─────
```

Now you give me problems to determine.

```
#   2 7     1 3
# - 1 8   - 7
    ─────   ─────
    1 9     1 6
    ─────   ─────
```

got the bug!

Please describe the bug.

student adds ten to answer.

here are some problems to test your hypothesis.

```
   2 5
 - 1 2
 ─────
#  2 3
 ─────
```

That is not the bug I have .... etc

Figure 18-2: A Typical BUGGY Session

18.4.2 A Model for Subtraction
How can the techniques of mathematical reasoning be used to build a model

of a student's subtraction procedure?

The model we will build is based, not on the BUGGY model, which is rather messy, but on the conceptually cleaner model of [O'Shea and Young 78]. We will define a two parameter predicate, *subtract*. Both parameters of subtract are terms representing subtraction sums. The first parameter will be a sum with an empty answer slot, and the second parameter will be the same sum, but with the answer filled in, e.g.

$$
\text{subtract} \left(\begin{array}{c} 4\ 8 \\ -1\ 9 \\ \hline \\ \hline \end{array} , \quad \begin{array}{c} 4\ 8 \\ -1\ 9 \\ \hline 2\ 9 \\ \hline \end{array} \right)
$$

subtract asserts that its second parameter is the 'answer' to the sum posed by its first. The scare quotes indicate that the 'answer' may be incorrect.

---

1) finished(Sum) → subtract(Sum,Sum)

2) ~finished(Sum1) & process-column(Sum1,Sum2) & shift-left(Sum2,Sum3) & subtract(Sum3,Sum4) → subtract(Sum1,Sum4)

3) subtrahend(Sum1) = minuend(Sum1) & result(0,Sum1,Sum2) → process-column(Sum1,Sum2)

4) subtrahend(Sum1) > minuend(Sum1) & add-ten-to-minuend(Sum1,Sum2) & decrement(Sum2,Sum3) & take-difference(Sum3,Sum4) → process-column(Sum1,Sum4)

5) subtrahend(Sum1) < minuend(Sum1) & take-difference(Sum1,Sum2) → process-column(Sum1,Sum2)

Figure 18-3: Clauses Defining subtract and process-column

subtract is defined by the clauses given in figure 18-3.

The meanings of the other predicates and functions in figure 18-3 are as follows:

- In a sum term, the column which must be worked on next, is marked. shift-left(Sum1,Sum2) asserts that Sum1 and Sum2 are identical, except that the mark is one column further left in Sum2.

- finished(Sum) asserts that the mark in Sum is beyond the leftmost column of numbers, i.e. the sum is complete.

- process-column(Sum1,Sum2) asserts that Sum1 and Sum2 are

identical, except that the marked answer slot of Sum1 is blank and that of Sum2 is 'correctly' filled in.

- subtrahend(Sum) is the bottom digit in the marked column of Sum.

- minuend(Sum) is the top digit in the marked column of Sum.

- result(Num,Sum1,Sum2) asserts that Sum1 and Sum2 are identical except that the marked answer slot of Sum1 is blank and that of Sum2 contains Num.

- add-ten-to-minuend(Sum1,Sum2) asserts that Sum1 and Sum2 are identical except that the marked minuend of Sum2 is 10 bigger than the marked minuend of Sum1.

- decrement(Sum1,Sum2) asserts that Sum1 and Sum2 are identical except that the minuend to the left of the marked one is 1 less in Sum2.

- take-difference(Sum1,Sum2) asserts that Sum1 and Sum2 are identical except that the marked answer of Sum1 is blank and that of Sum2 contains the *positive* difference between the minuend and the subtrahend.

Only subtract and process-column are defined in figure 18-3. Further clauses are needed to define the remaining predicates and functions.

Subtraction sums can be calculated by creating new goal clauses, e.g.

$$\text{subtract}(\begin{array}{r} 4\,8 \\ -\,1\,9 \\ \hline \end{array} , \quad \text{Answer}) \rightarrow$$

and resolving these with the above clauses. When the empty clause has been derived, Answer will be found to have had a term substituted for it representing the finished sum. A sample proof is outlined in figure 18-4. For the sake of conciseness, the sum terms in it have been written in a simpler format.

Consider what will happen if the 'decrement' literal of clause 4) is deleted and replaced by the literal Sum2 ≡ Sum3, i.e. Sum2 and Sum3 are identical.

Suppose the modified subtract were used to calculate the sum:

$$\begin{array}{r} 4\,8 \\ -\,1\,9 \\ \hline \\ \hline \end{array}$$

Since the minuend 4 would not be decremented during the resolving of the body of clause 4) the answer generated would be 39. This is a typical error produced by students. Another typical error is given in the following exercise.

Exercise 61: How could the clauses of figure 18-3 be modified to produce the following behaviour?

$$
\begin{array}{ccc}
37 & 65 & 27 \\
-19 & -23 & -64 \\
\hline
22 & 42 & 43 \\
\hline
\end{array}
$$

You may delete or replace any clause or proposition.

---

subtract(48−19=??, Answer) →

$\downarrow$ 2)

~finished(48−19=??) & process-column(48−19=??,Sum2) & shift-left(Sum2,Sum3) & subtract(Sum3,Answer) →

$\downarrow$

process-column(48−19=??,Sum2)& shift-left(Sum2,Sum3) & subtract(Sum3,Answer) →

$\downarrow$ 4)

subtrahend(48−19=??) = minuend(48−19=??) & add-ten-to-minuend(48−19=??,Sum2) & decrement(Sum2,Sum3) & take-difference(Sum3,Sum4) & shift-left(Sum4,Sum5) & subtract(Sum5,Answer) →

$\downarrow$

subtract(38−19=?9,Answer) →

$\downarrow$

Figure 18-4: An Outline of a subtract Proof

18.4.3 Psychological Validity

The BUGGY program is successful because the subtraction procedure it is based on appears to do subtraction problems in much the same way as human students do. When a computer program does something in the same way as a human, we say it is *psychologically valid.*

But we do not know what goes on in a person's head when he solves a problem, so how can we tell when a program is psychologically valid?

We can never be sure, but we can accumulate evidence. The sort of evidence which counts, is well illustrated by the subtraction program outlined in figure 18-3.

- The program should do the task in the same way as the human it is modelling. Since different humans will do the task in different ways we will usually have to content ourselves with a different program for each human.

- However, a program will earn more credit if it can be easily modified to account for the behaviour of different humans; or the same human on different days. Our subtraction program can be easily modified, by the deletion or replacement of a few literals and clauses, to account for correct subtraction behaviour and a wide range of commonly occurring erroneous behaviour.

- A program gains even more credit if it can be modified in steps to account for different stages in the development of the human's ability. Our subtraction program also has this property. See [O'Shea and Young 78] for details.

Psychologically validity is only established with respect to certain aspects of behaviour. We are not concerned to model the child staring out the window or poking his nose. But all scientific models have this character. Newton's laws of motion do not account for the colours of the objects in motion, or why they were thrown in the first place.

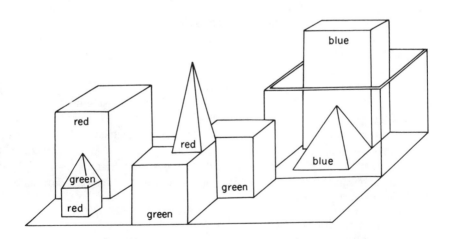

Figure 18-5: The Winograd Blocks World

18.5 Determining the Meaning of English Text

For our final application of mathematical reasoning techniques we turn to the area of natural language understanding programs. We will describe a technique developed by Terry Winograd [Winograd 72] for deciding which of several meanings to assign to an ambiguous sentence.

Winograd's program, SHRDLU, was able to hold a conversation about the world of toy blocks illustrated in figure 18-5.
It was able to obey commands like
"Put the blue pyramid on the block in the box."
Unfortunately, this sentence is ambiguous: is the blue pyramid on the block or is the block in the box?

SHRDLU discovered the ambiguity by using rules of grammar to break the sentence into parts. This can be done in two ways.

"Put (the blue pyramid on the block) in (the box)."
or
"Put (the blue pyramid) on (the block in the box)."

Each of the bits in parentheses is a noun phrase. SHRDLU worked on the principle that each definite noun phrase (roughly, one starting "the ...") should refer to something in the blocks world. Of the four definite noun phrases above, three were found to have such referents, but one did not. The offending phrase was "the blue pyramid on the block", so the first reading above was rejected and the second accepted.

To choose a referent for a phrase SHRDLU used an inference technique identical, in spirit, to that we have developed in this book. Each noun phrase was first translated into clauses, the object described by the noun phrase being represented by a variable. (Compare the 'Intermediate Representation' of chapter 14.) This clause was then resolved against similar clauses representing the blocks world. During the process of producing the empty clause a term was substituted for the noun phrase variable. This term represents the referent of the phrase. Thus this technique offers a partial solution to the noun phrase reference problem described in chapter 14.

In the example above the phrase 'the blue pyramid on the block' would be translated into the clause:

on (Thingy,block1) & colour(Thingy,blue) &
type(Thingy,pyramid) →   (i)

and the phrase 'the block in the box' into the clause:

in (Whatsit,block) & type(Whatsit,block) →   (ii)

The blocks world would be represented by a series of clauses.

$\rightarrow$ type(pyr1,pyramid) $\rightarrow$ type(block1,block) $\rightarrow$ type(box1,box)
$\rightarrow$ colour(pyr1,blue) $\rightarrow$ in(block1,box1)
....

The empty clause can be derived from clause (ii) and those representing the blocks world. The variable, Whatsit, being assigned the constant, 'block1'. However, the empty clause cannot be derived from (ii). Thus the first phrase has no referent.

Computer programs, like SHRDLU, which attempt to 'understand' written English, make heavy use of inference techniques. So too do programs for visual perception and speech perception. All cognitive tasks seem to involve processes of inference. Mathematics is an excellent domain for understanding such processes. The initial knowledge base is small and clear, as opposed to vision or speech where masses of messy data must first be extracted with a television camera or microphone. In Mathematics we can concentrate on the issues of representing knowledge and controlling search.

18.6 Logic Programming

The techniques of automatic inference, pattern matching, relational representation, etc are useful for building a wide variety of different kinds of computer program. A good way of making them more widely available is to provide them as primitives in a programming language. The resolution theorem provers, described in chapters 5 and 6, provide a basis for building programming languages with such primitives.

The idea that resolution theorem provers could be used as programming languages is called *Logic Programming*. It is due to Bob Kowalski, Alain Colmerauer, Pat Hayes and Cordell Green. The practical realization of it is a programming language called *PROLOG*, originally build by Colmerauer's group at Marseille, and developed at the universities of Edinburgh and Waterloo, Imperial College and several other places [Clocksin and Mellish 81]. PROLOG is a Lush Resolution theorem prover, with a depth first search strategy, and with the addition that some predicates are not satisfied by resolution, but are specially evaluated by the computer and then deleted. An example is the 'write' predicate, which causes its parameter to be printed on the users terminal and is then deleted from the clause. PROLOG programs are sets of Horn clauses defining a set of predicates. To run a program, the calculation to be made is defined as a goal clause, and 'proved' by Lush Resolution plus evaluation of the special predicates.

The BUGGY-type subtraction procedure defined in figure 18-3 is an example of a logic program. With slight syntactic changes this could be input to a computer running PROLOG, and used to calculate subtraction sums. Here is another example; for appending two lists.

1. → append(nil,List,List)

2. append(Cdr,List,Ans) →
append(cons(Car,Cdr), List,cons(Car,Ans))

Compare this with the definition of append given in section 11.2. We have replaced the two parameter function append with a three parameter predicate append. The effect of the explicit cond is obtained by using two clauses and the pattern of the first parameter.

To append the lists <one, thing>, <and, another> we first create the goal clause:

3. append(<one, thing>, <and , another>, Result) →

We then use this goal clause as the top clause of a Lush Resolution proof with axioms 1 and 2. During the proof the variable 'Result' will be bound to a term, and this term is the result of appending the two lists. The search tree is given in figure 18-6. There is, in fact, no search, i.e. the calculation is deterministic. During this proof the output parameter, Result, is bound to the term, <one, thing, and, another>, as required.

A characteristic of logic programs is that a procedure written to calculate one function, can often be used to calculate its inverse. For instance, the append procedure above can be used to split a list into a front and back sublists, by merely reversing the roles of input and output parameters, i.e. if PROLOG is given the goal

append(Front, Back, <one, thing, and, another>) →

then it will find five distinct proofs with output substitutions,

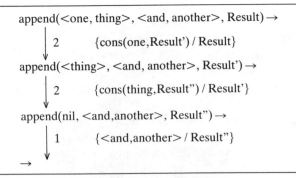

Figure 18-6: The Appending of Two Lists

{nil/Front, <one, thing, and, another>/Back}
{<one>/Front, <thing, and, another>/Back}
{<one, thing>/Front, <and, another>/Back}
{<one, thing, and>/Front, <another>/Back}   and
{<one, thing, and, another>/Front, nil/Back}

The search tree is given in figure 18-7.

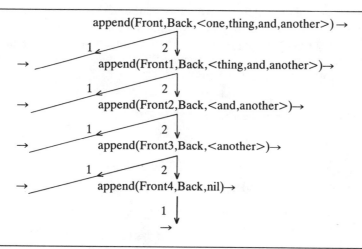

Figure 18-7: The Splitting of a List into Front and Back

Exercise 62: Draw the search tree generated by axioms 1 and 2 applied to the goal clause append(<one, thing>, Back, <one, thing, and, another>) →.

PROLOG is quite widely used as a programming language in Artificial Intelligence, particularly in modelling problem solving and natural language understanding. It has also been successful used as a first programming language for school children, and as a language for building database information systems. Its simple syntax and high level primitives make it an easy language for novices to learn and to use to build powerful programs quickly. Experienced programmers take a little longer, because they have first to overcome their prejudices about what a programming language should be like.

PROLOG is particularly good for building mathematical reasoning programs. Since Lush Resolution plus depth first search constitutes a very weak theorem prover, so it is best not to use it as it stands, but to build a more powerful theorem prover on top of it. This is relatively easy to do, as the built-in unification and relational representation provide just the primitives that are required. Some examples can be found in appendix I.

18.7 Summary

In this book we have seen how it is possible to build computer programs which model aspects of mathematical reasoning. We have addressed some of the problems involved, e.g. how can the search for a proof be guided so that a combinatorial explosion is avoided. We have discussed various techniques that can provide this guidance.

The techniques developed for modelling mathematical reasoning are finding increasing applications: providing mathematical aids and programming languages, helping to teach Mathematics and giving us a better understanding of cognitive processes. These applications are impinging on all our lives. To make best use of them and avoid abuse of them, we should all understand better how they work.

Further Reading Suggestions

Introductory accounts of:

- MACSYMA can be found in [Mathlab 77];

- the Wos/Overbeek/Lusk theorem prover can be found in [Wos 82];

- several tuition programs, built by Goldberg, can be found in [Goldberg & Suppes 74];

- BUGGY can be found in [Brown and Burton 78];

- SHRDLU can be found in [Winograd 72]; and

- PROLOG can be found in [Clocksin and Mellish 81].

Appendix I
Some Artificial Mathematicians Written in Prolog

In this appendix we turn some of the descriptions of artificial mathematicians, given in this book, into working computer programs. The programming language we have chosen for this task is the PROLOG language, mentioned in section 18.6. This is a natural choice since PROLOG is itself based on the early work on theorem proving, so this appendix serves two illustratory roles at once – an implementation of artificial mathematicians and an application of them to programming. A further advantage is that much of the groundwork, required for you to understand PROLOG programs, has already been laid in chapters 5 and 6 and section 18.6.

A Bluffer's Guide to PROLOG

A PROLOG program is a set of Horn clauses, but the notation differs slightly from the traditional notation, that we have used so far. In classical logic a Horn clause may be written,

$$P_1 \,\&\, ... \,\&\, P_n \to Q$$
$$\text{or}$$
$$P_1 \,\&\, ... \,\&\, P_n \to$$
$$\text{where } n \geqslant 0$$

In PROLOG these will be written:

$$Q :- P_1, ..., P_n.$$
$$\text{or}$$
$$:- P_1, ..., P_n.$$

that is: the antecedent is written to the right of the implication arrow; the consequent to the left of the arrow; the arrow itself is reversed and written as :– and the & signs are replaced by commas, with a full stop at the end. The idea of putting the consequent (Q) of the clause to the left of the antecedent (the P_is) is to emphasise that the antecedent constitutes the body of a procedure for calculating Q. In the case $n=0$, the clause '$Q :-.$' may be optionally written as 'Q.'. For the PROLOG clauses to be accepted by the computer, these conventions have to be strictly observed.

PROLOG permits a special syntax for lists. Lists are written as a sequence of terms separated by commas enclosed in square brackets, e.g. [this,is,a,list]. the empty list, nil, can be written [] and cons(a,b) can be written [a|b].* Thus in standard PROLOG format the procedure for appending two lists, given in section 18.6, is written as follows:

append([],List,List).

append([Car|Cdr],List,[Car|Ans]) :– append(Cdr,List,Ans).

A Straightforward Implementation

The most straightforward way to build artificial mathematicians in PROLOG is to use it as a theorem prover, representing the axioms of a theory and the conjectures directly as PROLOG clauses. For instance, the reflexive and twisted transitivity axioms of chapter 6 can be represented as:

equal(X,X).   (iii)

equal(U,W) :– equal(U,V), equal(W,V).   (iv)

and the conjectured symmetry axiom can be represented by the assertion

equal(x,y).   (v)

and goal clause

:– equal(y,x).   (vi)

To prove the conjecture, clauses (v), (iii) and (iv) are input to PROLOG, in that order, and the goal clause (vi) is called. PROLOG will then apply Lush Resolution with a depth first search strategy; clauses are selected in the order in which they were input and literals are selected left to right. This search strategy generates the following proof without deviation:

:– equal(y,x).
     (iv)

:– equal(y,V), equal(x,V).
     (iii)

:– equal(x,y).
     (v)

:–

*This is at variance with the notation used in the rest of this book, where angle brackets were used for lists and square brackets for bags.

However, the PROLOG search strategy is very simple minded and is not always so lucky in finding proofs. In particular it is very sensitive to the order in which clauses were input. If we had input clause (v) after clause (iv) above then the search would not have terminated.

Exercise 63: Check this by drawing the new search tree to depth 5.

To improve upon this we must stop using PROLOG *as* a theorem prover and instead use it as an implementation language for *building* theorem provers!

A Heuristic Search Theorem Prover

Suppose that we wanted to build a heuristic search theorem prover. Instead of representing each axiom directly as a PROLOG clause, we will need to step up to the meta-level (cf chapter 12); each axiom will be represented as a term occuring as the parameter of some predicate.* For instance, we may use the ternary predicate *is-clause*, which relates a clause's name, its consequent and its antecedent e.g.

is-clause(hypothesis, [equal(x,y)], []).

represents the clause → equal(x,y), and names it 'hypothesis'. The antecedent and consequent are represented as lists of propositions. The remaining clauses** in our example may be represented as:

** in our example may be represented as:

is-clause(reflexive, [equal(X,X)], []).

---

* We will have axioms like (iv), represented as clauses at the object-level, and theorem proving code, represented as clauses at the meta-level. This may be a source of notational confusion, so we will say 'PROLOG clause' when referring to the meta-level.

** Strictly speaking these are clause *schemata* rather than just clauses. This is because we have used PROLOG variables to represent object-level variables, rather than the PROLOG constants prescribed in chapter 12. We have done this so that we can use the built-in PROLOG unification procedure when resolving clauses rather than build our own. This trick can be legitimized by regarding U, V, X, etc as meta-variables ranging over object-level terms, and → equal(X,X), etc as clause schemata.

is-clause(twisted, [equal(U,W)], [equal(U,V), equal(W,V)]).

is-clause(hypothesis, [equal(x,y)], []).

is-clause(goal, [], [equal(y,x)]).

Our heuristic search theorem prover can be built by defining a proposition,
heuristic(Agenda)

which is true iff Agenda can form the agenda of a successful heuristic search
tree. An agenda is a list of elements of the form *pair(Score,Name)*, where
Name is the name of a clause and Score is the number which it is assigned by
the evaluation function. Agendas are ordered with the best scores first. We
will use the length of the clause as the evaluation function, so that lowest
scores are best. *heuristic* can be defined recursively.

It is true if Agenda contains the empty clause.

heuristic(Agenda) :-
    member(pair(0,Empty),Agenda).

member(E,L) is true iff E is a member of the list L. When called this clause
checks to see whether the Agenda of the heuristic search tree contains a
clause whose score is 0.

It is also true if Agenda can be extended, by the rules of heuristic search,
into the Agenda of a successful heuristic search tree.

heuristic([pair(Score,Current) | Rest]) :-
    setof1(Clause,successor(Current,Clause),NewClauses),
    add-to-agenda(NewClauses,Rest,NewAgenda),
    heuristic(NewAgenda).

setof1(X,P(X),Set) is our version of the set formation operation,
Set = {X : P(X)}. Thus NewClauses is the set of all objects, Clause,
which bear the relation successor(Current,Clause) to Current.
successor(Current,Clause) means that Clause is a resolvant, one of whose
parents is Current. *add-to-agenda(NewClauses,Rest,NewAgenda)* is true
iff NewAgenda is the agenda formed from Rest by inserting each of the
clauses in NewClauses together with their scores.

When called, this PROLOG clause picks one of the undeveloped clauses
with the best score (Current), finds all it successors (NewClauses), replaces
Current by NewClauses and then calls *heuristic* recursively.

member is easily defined using append.

member(E,L) :– append(L1,[E|L2],L).

setof1 is very similar to the PROLOG evaluable predicate, *setof;* the only difference being that *setof(X,P(X),Set)* fails if $\sim \exists X\ P(X)$, instead of instantiating Set to []. *setof1* can be defined as follows:

setof1(X,P,Set) :– setof(X,P,Set), !.

setof1(X,P,[]).

The !, in the first clause, is a special evaluable 'predicate' in PROLOG whose side-effect is to prune the PROLOG search tree of any remaining choice points for this call of *setof1*. It prevents later failures from causing back up to the second clause.

It only remains to define *successor* and *add-to-agenda* to complete our theorem prover.

We start with *successor*.

successor(Current,Clause) :-
        factor(Current,Clause).

successor(Current,Clause) :-
        is-clause(Parent,—,—),
        (resolve(Current,Parent,Clause) ;
            resolve(Parent,Current,Clause)).

factor(C,F) is true iff F is a factor of clause C. *resolve(P1,P2,R)* is true iff R is a resolvant whose parents are P1 and P2, and where the proposition resolved on occurs in the consequent of P1 and the antecedent of P2. Semi-colon is the built-in PROLOG version of disjunction, so the definition of successor above is not in clausal form, but packs two clauses together. Each successive call of this definition picks a potential second parent (Parent) and tries resolving it with Current in each of the two possible ways, to form a new resolvant (Clause).

Exercise 64: Modify the theorem prover so that it obeys the input restriction – one parent of each resolvant is always an input clause. [Hint: Why is this exercise just here?]

A successful call of resolve will return the name of a new clause. This name is not much use unless it is also the first parameter of an *is-clause* assertion, e.g. a call of

resolve(goal,twisted,Clause).

may succeed with substitution {resolvant1/Clause} (say), but it must also input a new PROLOG clause of the form:

is-clause(resolvant1,[], [equal(y,V'), equal(x,V')]).   (vii) *resolve* can be defined by the clause:

resolve(Parent1, Parent2, Resolvant) :-
        is-clause(Parent1, Consequent1, Antecedent1),
        is-clause(Parent2, Consequent2, Antecedent2),
        select(Proposition, Consequent1, RestConse1),
        select(Proposition, Antecedent2, RestAnte2),
        append(RestConse1, Consequent2, Consequent),
        append(Antecedent1, RestAnte2, Antecedent),
        gensym(resolvant,Resolvant),
        assert(is-clause(Resolvant,Consequent,Antecedent)).

where *select(E,L,R)* is true iff L is the list formed by inserting E into the list R. The two calls of *select* have the effect of selecting and unifying two propositions, one from the the consequent of the first parent and one from the antecedent of the second parent. Note how the two propositions are unified - and the unifier applied to the remaining propositions – by built-in, PROLOG unification.* The two calls of *append* have the effect of making new lists for the consequent and antecedent of the resolvant by appending together the remnants of the consequents and antecedents of the parents.

select is easily defined using append.

select(E,L,R) :– append(L1,[E|L2],L), append(L1,L2,R).

Both *gensym* and *assert* take us outside of Predicate Logic. Neither can be sensibly interpreted as bona fide, first order predicates. *assert* is an evaluable 'predicate' of PROLOG. When called it always succeeds with the side effect of inputing its parameter as a new PROLOG clause, e.g. (vii) above. The opposite of *assert* is *retract*, it succeeds iff it can delete a PROLOG clause matching its parameter.

factor can be defined by the clause:

factor(Clause, Factor) :-
        is-clause(Clause,Consequent,Antecedent),
        select(Proposition,Consequent,OneGone),
        select(Proposition,OneGone,TwoGone),
        gensym(factor,Factor),
        assert(is-clause(Factor,OneGone,Antecedent)).

factor(Clause, Factor) :-
        is-clause(Clause,Consequent,Antecedent),

* In fact, this use of PROLOG unification makes our theorem prover unsound ! The PROLOG unifier omits the occurs check (see chapter 17). Thus it will unify equal(X,X) and equal(Y,Y+1).

```
select(Proposition,Antecedent,OneGone),
select(Proposition,OneGone,TwoGone),
gensym(factor,Factor),
assert(is-clause(Factor,Consequent,OneGone)).
```

The first PROLOG clause tries to find two occurrences of the same proposition in the consequent and the second PROLOG clause does the same thing in the antecedent. *select* is used to find the two occurrences, just one of them being omitted in the definition of the new factor.

gensym must be defined by us. Its role is to invent names for new clauses, e.g. resolvant1. It is always called with its first parameter a constant and its second parameter a variable; the call instantiates the second parameter to a constant composed of the first parameter and a number. Each time gensym is called the number is incremented, giving a new constant for the second parameter. When '*resolvant*' is the first parameter, as it is in resolve, gensym will generate the sequence of constants: resolvant1, resolvant2, resolvant3, ... as the second parameter.

The definition of *gensym* is:

```
gensym(Prefix,Var) :-
        var(Var), atomic(Prefix),
        get(Prefix,N),
        N1 is N+1,
        assert(latest(Prefix,N1)),
        concat(Prefix,N1,Var).

get(Prefix,N) :- retract(latest(Prefix,N)), !.
get(Prefix,0).

concat(N1,N2,N) :-
        name(N1,Ls1),
        name(N2,Ls2),
        append(Ls1,Ls2,Ls),
        name(N,Ls).
```

var(Var) is true iff Var is a PROLOG variable. *atomic(Prefix)* is true iff Prefix is a PROLOG constant. *N1 is N+1* is true iff N and N1 are both numbers and N1 is the sum of N and 1. *concat(Prefix,N1,Var)* is true iff Prefix, N1 and Var are all character strings and Var is the concatenation of Prefix and N1. *name(N,L)* is true iff N is a character string and L is a list of the ASCII code numbers of the characters in N. *get* cannot be assigned a meaning as a bone fide, first order predicate.

gensym works by asserting and retracting PROLOG clauses of the form 'latest(Prefix,N)' where N is the suffix of the last constant, with prefix Prefix, that it invented. It first checks that its parameters are of the appropriate types by calling *atomic(Prefix)*, which checks that Prefix is a constant, and *var(Var)*, which checks that Var is a variable. Then *get(Prefix,N)* recovers the last suffix number, N, used for Prefix, if there was one, or 0 otherwise. The role of the ! in the first clause of get is to prevent later failures from remaking the instantiation of N. Having found the value of N, 1 is added to it to form N1, a new 'latest' clause is asserted, and Prefix and N1 are concatenated together to form the new constant, Var.

concat(Prefix,N1,Var) works by breaking Prefix and N1 into a list of character numbers using *name*, appending these lists together and then using *name* in the other direction to make a constant out of this new list of character numbers. *atomic, var, is, +* and *name* are all evaluable predicates or functions of PROLOG.

We now define *add-to-agenda(NewClauses, Old Agenda, New Agenda)*. It is defined recursively on the list structure of its first parameter.

add-to-agenda([],Agenda,Agenda).

add-to-agenda([Name|Rest],Agenda,NewAgenda) :-
        evaluate(Name,Score),
        insert-into-agenda(Agenda,Score,Name,MidAgenda),
        add-to-agenda(Rest,MidAgenda,NewAgenda).

evaluate(Name,Score) is true iff Score is the length of the clause called Name. *insert-into-agenda(Agenda,Score,Name,MidAgenda)* is true iff MidAgenda is the agenda formed by inserting pair(Score,Name) into the appropriate place in Agenda.

If NewClauses is the empty list then the agenda is not changed. Otherwise, the score of the first clause is worked out, the name and score are entered in the agenda and the rest of NewClauses are recursively inserted.

evaluate(Name,Score) is easily defined as:

evaluate(Name,Score) :-
        is-clause(Name,Consequence,Antecedent),
        length(Consequence,C),
        length(Antecedent,A),
        Score is C+A.

length(L,N) is true iff N is the length of the list L. The consequence and antecedent of the clause called Name are recovered, their lengths are found and added together. *length* is a primitive provided by PROLOG, so *evaluate* is now completely defined.

insert-into-agenda(OldAgenda,Score,Name,NewAgenda) is also defined recursively on its first parameter.

    insert-into-agenda([],Score,Name,[pair(Score,Name)]).

    insert-into-agenda([pair(Score1,Name1)|Rest],Score,Name,
                [pair(Score,Name),pair(Score1,Name1)|Rest]) :-
        Score =<Score1,
        !.

    insert-into-agenda([X|Rest],Score,Name,[X|NewRest]) :-
        insert-into-agenda(Rest,Score,Name,NewRest).

Score =< Score1 is true iff the number Score is less than or equal to the number Score1.

If OldAgenda is the empty list then the new agenda is a singleton consisting of pair(Score,Name). Otherwise, there are two cases:

1. If Score is smaller than the score of the first element on OldAgenda (Score1) then we put the new pair on the front of OldAgenda to form NewAgenda.

2. Otherwise, we insert the new pair recursively into the rest of OldAgenda (Rest).

The use of !, in the clause defining the first case, prevents PROLOG from incorrectly backing up onto the second case. =< is a PROLOG evaluable predicate, so *insert-into-agenda* is completely defined, and this completes the definition of *heuristic*.

To run the heuristic search theorem prover it only remains to select a goal clause, e.g. goal above, and to make an agenda item as the first parameter of *heuristic*, e.g. [pair(1,goal)]. We now type

    :- heuristic([pair(1,goal)]).

to PROLOG.

Exercise 65: Type this program into a computer file and run it on the example clauses given above.

Exercise 66: Insert print messages into the program so it outputs a suitable commentary on its progress.

Building-in—Semantic Checking

In this section we will consider giving our theorem prover the ability to use models to prune the search tree à la Gelernter (see chapter 10). We call this semantic checking. This requires a new predicate *vet*, such that *vet(C1,C,I)* is true iff C is a variable free instance of clause C1, and C is false in the interpretation I. We must modify *successor* so that *successor(Current,Clause)* is true iff Clause is derived from Current by resolution or factoring *followed by vetting*. This can be done by making the previous definition of successor be a definition of successor1, and defining successor as:

    successor(Current,Clause) :-
      successor1(Current,Clause1),
      vet(Clause1,Clause,model1).

where model1 is the name of a model of the axioms and hypothesis of the conjecture. *vet* can be defined as follows:

    vet(Clause1,Clause,Interp) :-
        is-clause(Clause1,Conse,Ante),
        constants(Consts),
        checklist(instantiate(Consts),Conse),
        checklist(instantiate(Consts),Ante),
        false-clause(Conse,Ante,Interp),
        gensym(instance,Clause),
        assert(is-clause(Clause,Conse,Ante)).

constants(Consts) is true iff Consts is a list of constants. *checklist(P(X),YList)* is the PROLOG version of bounded universal quantification $\forall Y \epsilon YList\ P(X,Y)$. *instantiate(Consts,Prop)* is true iff Prop is a variable free proposition whose constants are all members of Consts. *false-clause(Conse,Ante,Interp)* is true iff Conse is a list of variable free propositions which are false in interpretation Interp and Ante is a list of variable free propositions which are true in Interp.

To *vet* a clause named Clause1 in an interpretation Interp the following steps are taken.

1. Its consequent, Conse, and antecedent, Ante, are looked up with

is-clause and the constants are looked up with *constants*.

2. Any variables in Conse and Ante are instantiated with constants from Consts.

3. The instantiated Conse and Ante are tested in Interp to see if they would form a false clause.

4. If so, then a new clause name, Clause, is invented by gensym and an appropriate is-clause is input to PROLOG.

constants is defined by a single clause, e.g.

    constants([x,y,z]).

This definition will vary according to the conjecture whose proof is sought. We have included all constants which appear in the clauses given in our equality example plus one more to appear in an axiom below.

instantiate can be defined as follows.

    instantiate(Consts,Constant) :-
            atomic(Constant).
    instantiate(Consts,Variable) :-
            var(Variable), member(Variable,Consts).
    instantiate(Consts,Complex) :-
            \+ atomic(Complex),
            \+ var(Complex),
            Complex =.. [Sym Paras],
            checklist(instantiate(Consts),Paras).

Both *not* and =.. are PROLOG evaluable predicates which take us outside Predicate Logic. \+ P is the PROLOG version of negation, it is true iff PROLOG cannot prove P.* *Complex =.. [Sym/Paras]* is true iff Complex is a complex term or formula, Sym is its function or predicate symbol and Paras is a list of its parameters.

instantiate(Consts,Expr) is defined by recursion on its second parameter, Expr. The first clause deals with the case where Expr is a constant, in which case there is nothing to do. The second clause deals with the key case where Expr is a variable. In this case the variable is instantiated to be a member of Consts. Otherwise, if Expr is complex, it is broken apart by =.. and instantiate is recursively applied to its parameters.

checklist(P,YList) can be defined recursively on its second parameter YList.

    checklist(P,[]) :- !.

*In some implementations \+ is available as *not* or *thnot*.

```
checklist(P,[Y|YList]) :– !,
     P =.. [Sym | XList],
     append(XList,[Y],Paras),
     Q =.. [Sym | Paras],
     Q,
     checklist(P,YList).
```

When YList is the empty list the call succeeds. Otherwise, P is applied to the first member of the list, Y, and checklist recurses on the rest of the list. To apply P to Y it is first unpacked by =.. to reveal its predicate symbol, Sym, and any partially applied parameters, XList. Y is appended to the end of XList and the result repacked with Sym using =.. to form Q. Q is then called as a PROLOG procedure.

false-clause is defined as follows.

```
false-clause(Consequent,Antecedent,Interp) :-
     checklist(meaning(Interp,false), Consequent),
     checklist(meaning(Interp,true), Antecedent).
```

meaning(Interp, Value, Expr) is true iff Value is the meaning in interpretation, Interp, of expression, Expr. *false-clause* uses *checklist* to check that each member of the consequent is false in Interp and each member of the antecedent is true in Interp.

meaning(Interp,Value,Expr) can be defined recursively on the expression tree structure of its third parameter, Expr.

```
meaning( Interp, Value, Constant ) :-
     atomic(Constant), !, interpret(Interp, Value, Constant).
```

```
meaning( Interp, Value, Complex ) :-
     Complex =.. [Sym | Paras], !,
     maplist(meaning(Interp), Vals, Paras),
     Complex1 =.. [Sym | Vals],
     interpret(Interp, Value, Complex1).
```

interpret(Interp, Value, Expr) is similar to *meaning(Interp, Value, Expr)* except that Expr is either an element of the universe of Interp or a function or predicate whose parameters are elements of the universe of Interp.

The PROLOG clauses defining *meaning* deal with the following two cases:

1. If Expr is a constant then *interpret* is used to find the assigned member of the universe of Interp.
2. If Expr is a complex expression then: it is broken apart with =..; the

meanings of each of its parameters is recursively calculated using *maplist*; these are applied to its dominant symbol by $=..$; and its value is calculated with interpret.

maplist(P, YList, ZList) can be defined recursively on its second parameter YList in an analogous way to *checklist*.

maplist(P,[],[]) :– !.

maplist(P,[Y|YList],[Z|ZList]) :– !,
        P =.. [Sym | XList],
        append(XList,[Y,Z],Paras),
        Q =.. [Sym | Paras],
        Q,
        maplist(P,YList,ZList).

The definition of interpret depends on the interpretation we wish to define. The clauses below define the interpretation, *model1*, which is suitable for use with the equality example from the last section. We must assign meanings to the *constants x, y and z* and the predicate *equal*.

interpret(model1, 2, x).

interpret(model1, 2, y).

interpret(model1, 3, z).

interpret(model1, true, equal(X,Y)) :-
        X == Y.

interpret(model1, false, equal(X,Y)) :-
        X \== Y.

The assignment to the constants is straightforward, but the assignment to *equal* deserves comment. It uses the PROLOG evaluable predicates $X==Y$ and $X\backslash==Y$ which are true iff X and Y are strictly identical or strictly non-identical, respectively, i.e. two numbers are *equal* iff they are the same number. These are used to define a calculation procedure for *equal*.

This completes the definition of *vet*, and the heuristic search theorem prover with semantic checking can now be run. However, it will not do any interesting pruning on our current equality example since this is too trivial to generate any prunable clauses. To see the benefit we will need to add some more axioms for instance the 'funny' axiom:

is-clause(funny,[equal(X,Y)],[equal(X,z),equal(z,Y)]).

When resolved with clause 'goal' this will generate

is-clause(instance1,[],[equal(x,z),equal(z,y)]).

But equal(x,z) has meaning false in model1, so the clause is true and fails the vetting, as required.

> *Exercise 67:* Type in the above clauses and run the theorem prover on the equality example.

> *Exercise 68:* Insert a loop check to insure that the same clause does not get added to the data base twice.

> *Exercise 69:* Represent, using *is-clause*, the clauses from the 'not divides' example of section 10.5. Represent, using *interpret*, the models *arith2* and *arith3*. Run the theorem prover on this example.

The Complete Program

The entire program is repeated below, including the answers to some of the exercises. Text between /* and */ and after % is comment.

/* HEURISTIC SEARCH THEOREM PROVER */

go :– heuristic([pair(1,goal)]).

/* Top Level Stuff */
heuristic(Agenda) :-
        member(pair(0,Empty),Agenda).

heuristic([pair(Score,Current) | Rest]) :-
        setof1(Clause,successor(Current,Clause),NewClauses),
        add-to-agenda(NewClauses,Rest,NewAgenda),
        heuristic(NewAgenda).

successor(Current,Clause) :-
        successor1(Current,Clause1),
        vet(Clause1,Clause,model1),
        message(Clause).   %new

successor1(Current,Clause) :-
        factor(Current,Clause).

successor1(Current,Clause) :-
        is-clause(Parent,—,—),
          resolve(Current,Parent,Clause) ;
             resolve(Parent,Current,Clause)).

```
/* Rules of Inference */
resolve(Parent1, Parent2, Resolvant) :-
        is-clause(Parent1, Consequent1, Antecedent1),
        is-clause(Parent2, Consequent2, Antecedent2),
        select(Proposition, Consequent1, RestConse1),
        select(Proposition, Antecedent2, RestAnte2),
        append(RestConse1, Consequent2, Consequent),
        append(Antecedent1, RestAnte2, Antecedent),
        gensym(resolvant,Resolvant),
        assert(is-clause(Resolvant,Consequent,Antecedent)).

factor(Clause, Factor ) :-
        is-clause(Clause,Consequent,Antecedent),
        select(Proposition,Consequent,OneGone),
        select(Proposition,OneGone,TwoGone),
        gensym(factor,Factor),
        assert(is-clause(Factor,OneGone,Antecedent)).

factor(Clause, Factor ) :-
        is-clause(Clause,Consequent,Antecedent),
        select(Proposition,Antecedent,OneGone),
        select(Proposition,OneGone,TwoGone),
        gensym(factor,Factor),
        assert(is-clause(Factor,Consequent,OneGone )).

/* Print Message */            %new
message(Clause) :-
        is-clause(Clause,Conse,Ante),
        write('New Clause, '),
        write(Clause),
        write(', is '),
        write(Ante),
        write(' → '),
        write(Conse),
        write('
').

/* Evaluation Function */
add-to-agenda([],Agenda,Agenda).

add-to-agenda([Name | Rest],Agenda,NewAgenda) :-
        \+ in(Name),            %new
        evaluate(Name,Score),
        insert-into-agenda(Agenda,Score,Name,MidAgenda).
```

```
                add-to-agenda(Rest,MidAgenda,NewAgenda).

        evaluate(Name,Score) :-
                is-clause(Name,Consequence,Antecedent),
                length(Consequence,C),
                length(Antecedent,A),
                Score is C+A.

        insert-into-agenda([],Score,Name,[pair(Score,Name)]).
        insert-into-agenda([pair(Score1,Name1) | Rest],Score,Name,
        [pair(Score,Name),pair(Score1,Name1)| Rest]) :-
                Score =< Score1,
                !.


        insert-into-agenda([X|Rest],Score,Name,[X|NewRest]):-
                insert-into-agenda(Rest,Score,Name,NewRest).

        /* Loop Check */          %new
        in(Clause) :-
                is-clause(Clause,Conse,Ante),
                is-clause(Another,Conse,Ante),
                Clause \ == Another.

        /* SEMANTIC CHECKING */

        vet(Clause1,Clause,Interp):-
                is-clause(Clause1,Conse,Ante),
                constants(Consts),
                checklist(instantiate(Consts),Conse),
                checklist(instantiate(Consts),Ante),
                false-clause(Conse,Ante,Interp),
                gensym(instance,Clause),
                assert(is-clause(Clause,Conse,Ante)).

        instantiate(Consts,Constant) :-
                atomic(Constant).

        instantiate(Consts,Variable) :-
                var(Variable), member(Variable,Consts).

        instantiate(Consts,Complex) :-
                \ + atomic(Complex),
                \ + var(Complex),
                Complex =.. [Sym|Paras],
                checklist(instantiate(Consts),Paras).
```

```
false-clause(Consequent,Antecedent,Interp) :-
        checklist(meaning(Interp,false), Consequent),
        checklist(meaning(Interp,true), Antecedent).

meaning( Interp, Value, Constant ) :-
        atomic(Constant), !, interpret(Interp, Value, Constant)..

meaning( Interp, Value, Complex ) :-
        Complex =.. [Sym | Paras], !,
        maplist(meaning(Interp), Vals, Paras),
        Complex1 =.. [Sym | Vals],
        interpret(Interp, Value, Complex1).
```

/* GENERAL UTILITIES */

/* List Processing */
```
    append([],List,List).
    append([Car|Cdr],List,[Car|Ans]) :- append(Cdr,List,Ans).
    member(E,L) :- append(L1,[E|L2],L).
    select(E,L,R) :- append(L1,[E|L2],L), append(L1,L2,R).

    /* Logical */
    setof1(X,P,Set) :- setof(X,P,Set), !.
    setof1(X,P,[]).
    checklist(P,[]) :- !.

    checklist(P,[Y|YList]) :- !,
                P =.. [Sym | XList],
                append(XList,[Y],Paras),
                Q =.. [Sym | Paras],
                Q,
                checklist(P,YList).

    maplist(P,[],[]) :- !.
    maplist(P,[Y|YList],[Z|ZList]) :- !,
                P =.. [Sym | XList],
                append(XList,[Y,Z],Paras),
                Q =.. [Sym | Paras],
                Q,
                maplist(P,YList,ZList).
```

/* Generate New Name */
```
    gensym(Prefix,Var) :-
            var(Var), atomic(Prefix),
```

```
            get(Prefix,N),
            N1 is N+1,
            assert(latest(Prefix,N1)),
            concat(Prefix,N1,Var).

get(Prefix,N) :- retract(latest(Prefix,N)), !.
            get(Prefix,0).

concat(N1,N2,N) :-
            name(N1,Ls1),
            name(N2,Ls2),
            append(Ls1,Ls2,Ls),
            name(N,Ls).


    /* EXAMPLE SPECIFIC STUFF */

    /* Axioms and Negated Conjecture */
    is-clause(reflexive, [equal(X,X)], [] ).
    is-clause(funny,[equal(X,Y)],[equal(X,z),equal(z,Y)]).
    is-clause(twisted, [equal(U,W)], [equal(U,V), equal(W,V)] ).
    is-clause(hypothesis, [equal(x,y)], [] ).
    is-clause(goal, [], [equal(y,x)] ).

    /* Constants */
    constants([x,y,z]).

    /* Interpretation Model1 */
    interpret(model1, 2, x).
    interpret(model1, 2, y).
    interpret(model1, 3, z).
    interpret(model1, true, equal(X,Y)) :-
            X == Y.
    interpret(model1, false, equal(X,Y)) :-
            X \== Y.
```

Further Reading Suggestions

More example of PROLOG programs can be found in [Coelho et al 80].

Appendix II
The Language of Trees

In developing computational models of mathematical reasoning we will have frequent recourse to *trees* as a descriptive device. We will use them to: describe expressions; the meaning of expressions and the search for a proof. It will help us in the the future if we define now what we mean by a tree and develop some notation for discussing them.

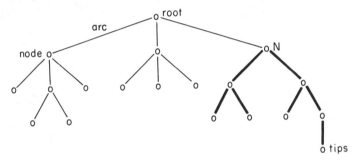

Figure 18-8: A Tree

Figure 18-8 is a drawing of a tree. It consists of some *nodes* (the os) and some *arcs* (the lines between them). The node at the top is called the *root* (our trees are Australian!). All nodes, except the root, have a unique *parent** node. These are the ones immediately above them. Some nodes have *daughter* nodes, immediately below them. Those which do not are called *tips*. The set of all daughters, daughters of daughters etc of a node are called its *descendants* The subtree consisting of all descendants of a node and the arcs between them is said to be *dominated* by the node. The subtree dominated by the node labelled N in figure 18-8 is printed in heavy type. Removing some nodes and arcs from a tree to form a subtree is called *pruning*.

A sequence of consecutive arcs and nodes is called a *path* and a path running from the root to a tip without repetition is called a *branch*. The *length* of a path is the number of arcs it contains. The length of the path from the root to a node is called its *depth* and the length of the path from a node to the farthest tip it dominates is called its *height*. The depth of the node N in figure 18-8 is 1 and its height is 3.

*There is a conscious analogy with family trees

Having introduced this notation once we will use it freely to describe: expression trees; semantic trees and search trees.

Appendix III
Alternative Notation

Predicate logic is a mathematical theory which remains essentially the same when expressed in a wide variety of notations, just as the theory of Arithmetic is unaltered by being expressed in Arabic or Roman numerals. However, just as arithmetic operations are easier to do in some numeral systems, so logical operations are easier to do in some notational systems. In this appendix we show how to express the same formulae in a wide variety of notations. As running examples we will use the binary predicate p and the ternary predicate q.

In the notation of this book, which is called Functional Form, the applications of p to the parameters a and b, and q to the parameters c, d and e, are represented by,

p(a,b) and q(c,d,e)

Note that the opening bracket comes after the predicate symbol.

A common variant: as used, for instance, in the programming language LISP; is to put the bracket in front of the predicate symbol.

(p,a,b) and (q,c,d,e)

This is called *Cambridge Polish*. It is an especially convenient notation if you want your pattern matcher not to distinguish between predicates and their parameters, e.g. a second order matcher.

Another variant on this theme is *Reverse Polish*. Here there are no brackets at all and parameters appear before the predicate, e.g.

a,b,p and c,d,e,q

The notations encountered so far form a natural family, which I shall call the *Polish* family. We will next consider a radically different family: the *Semantic Nets*. These are graphs (or networks) of nodes and arcs, representing a conjunction of assertions.

The simplest form is when each arc represents a predicate and each node a constant, e.g.

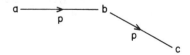

which in Functional Form would be expressed as p(a,b) & p(b,c). To represent a ternary (or greater arity) predicate there are various devices. The simplest format is

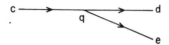

but this is fairly rare.

An alternative is to use a *Case Frame* formalism. In this notation nodes are used to represent, not only parameters, but also predicates *and* propositions. The arcs point from propositions to both parameters and predicates, and indicate which role these parameters and predicates play the proposition, e.g.

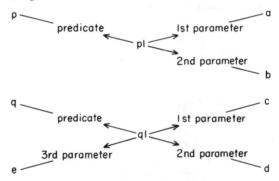

p1 and q1 stand for the propositions p(a,b) and q(c,d,e), respectively. The labels on the arcs are usually more mnemonic than 1st parameter, predicate, etc: they tend to indicate the type or role played by the parameter, e.g. agent, instrument, time, homeport, etc.

Many kinds of expression are difficult to represent in Semantic Net systems, for instance, negation, disjunction, quantification and the distinction between functions and predicates. Nevertheless, it can be done. The interested reader is referred to [Schubert 76].

The final family of notational systems we will consider are the *Frame* systems (also called schema). These are similar, in content, to the Semantic Net systems, but different in layout. One of the nodes is taken as the title of a table, the arcs leading from this node are listed in a column marked *slots* and the nodes they lead to are listed in a column marked *fillers*. We may chose to group arcs with the same label in a single row (see over page).

This Frame represents the formula

p(a,b) & p(a,c) & r(a,d)

a	slots	fillers
	p	b,c
	r	d

This notation is essentially the tabular form used by AM to represent the properties of a concept in chapter 13.

If we model our Frame on the Case Frame version of Semantic Nets then the title of the table is a particular proposition, the slots are the role names, 1st parameter, etc, and the fillers are the actual first parameter, predicate etc, e.g.

q1	slots	fillers
	predicate	q
	1st parameter	c
	2nd parameter	d
	3rd parameter	e

Again the role names are usually more mnemonic.

The various alternative notational systems are summarized in figure 18-9.

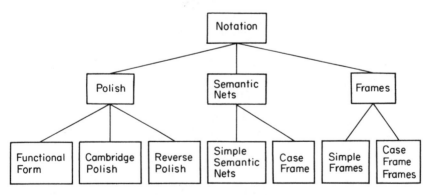

Figure III-1: Type Tree of Notations

Each of these notations present different advantages.

 – The Polish notations are more concise, especially when quantifiers, disjunction, etc are to be represented.

 – The Semantic Net notations suggest indexing the database of formulae on the parameters rather than the predicates.* The notations with role names allow extra information to be stored with them, e.g. type information or axioms.

Many computational inference systems use one of the above notations to represent theories which are not Predicate Logic. For instance, the system may include 'complete information' assumptions which violate the semantics of Predicate Logic, or they may include non-deductive processes like 'random associative walks' around the network. In section 4.3.5 we investigated the dangers of an inference system with no clear semantics.

*Most modern databases do both.

Appendix IV
Solutions to the Exercises

Chapter 2

1.

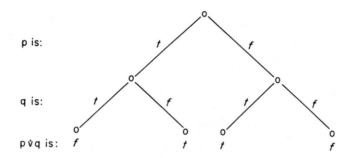

p is:

q is:

p v̇ q is: *f*   *t*   *t*   *f*

Semantic Tree for Exclusive Or

2.

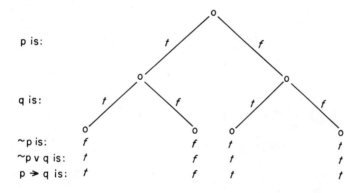

p is:

q is:

~p is: *f*   *f*   *t*   *t*
~p v q is: *t*   *f*   *t*   *t*
p → q is: *t*   *f*   *t*   *t*

Semantic Tree for ~p v q

Note that ~p v q and p → q have the same truth value for every assignment
to p and q.

3.

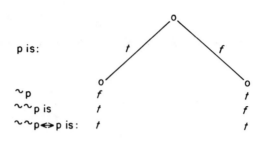

p is:	t	f
~p	f	t
~~p is	t	f
~~p ⟷ p is:	t	t

~~p ⟷ p is a Tautology

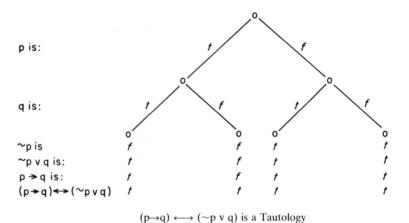

p is:	t			f	
q is:	t	f	t	f	
~p is	f	f	t	t
~p ∨ q is:	t	f	t	t
p → q is:	t	f	t	t
(p→q)⟷(~p∨q)	t	t	t	t

(p→q) ⟷ (~p ∨ q) is a Tautology

4.

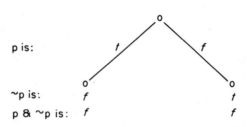

p is:	t	f
~p is:	f	t
p & ~p is:	f	f

p & ~p is a Contradiction

5. The argument can be formalized as:

{P → Q) & P} → Q

The semantic tree of this formula is:

    P is:

      Q is:

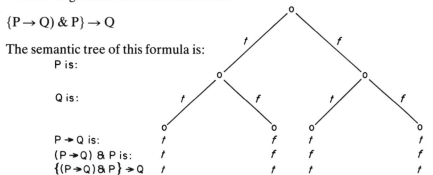

	t	f	t	f
P → Q is:	t	f	t	t
(P → Q) & P is:	t	f	f	f
{(P → Q) & P} → Q	t	t	t	t

which shows that it is a tautology, and hence that the argument form is correct.

Chapter 3

6.

    Continuous

$$\forall X \; \forall \epsilon \; \exists D \; \forall Y \quad |Y-X| < D \rightarrow |e^{\,Y} - e^X| < \epsilon$$

    Uniformly Continuous

$$\forall \epsilon \; \exists D \; \forall X \; \forall Y \quad |Y-X| < D \rightarrow |e^Y - e^X| < \epsilon$$

Note that the only difference between them is the order of the quantifiers, in particular, whether D is allowed to depend on X or to be uniform for all X.

7. The Darii syllogism can be formalized as:

$$\frac{\{\forall Y \; p(Y) \rightarrow q(Y)\} \; \& \; p(x)}{q(x)}$$

Let M be a typical model of the hypothesis. Hence, by the semantic tree for &, both $\forall Y \; p(Y) \rightarrow q(Y)$ and p(x) are true in M. $p(x) \rightarrow q(x)$ is an instance of $p(Y) \rightarrow q(Y)$ so, by the rule for $\forall$, it is true in M. Therefore, by the semantic tree for →, q(x) is true in M. Therefore, every model of the hypothesis is a model of the conclusion, and the argument form is correct.

8. Let M be a typical model of $\sim\forall X \; A(X)$. $\forall X \; A(X)$ is false in M. Therefore, by the rule of $\forall$, some variable free instance of A(X) is false in M. Therefore, some variable free instance of $\sim A(X)$ is true in M. Therefore, by the rule for $\exists$, $\exists X \sim A(X)$ is true in M.

Chapter 4

9. See section 8.1.

10. See section 9.2.3.

11.

$$X^0 = X$$

$$X^{s(Y)} = X.X^Y$$

Chapter 5

12. Let M be a model of C' v P and C" v ~P. Suppose that C' v C" is false in M. By the semantic tree for v, both C' and C" are false in M. Hence, by the semantic tree for v, both P and ~P are true in M – contradiction! Therefore, C' v C must be true in M. Therefore, C' v C is a logical consequence of C' v P and C" v ~P.

13.

1. q v ~r v s

2. q(f(f(Y))) v ~r(Y)

3. p(f(a),Y) v q(f(a),Y) v r(a)
    p(X,a) v q(X,a) v r(a)
    q(f(a),a) v r(a)

14. All but ~0=s(X), which can easily be written as 0=s(X) →, and the induction axiom, which is second order and not in clausal form.

15. Same as last exercise – but both these counterexamples can be rewritten as Horn clauses. For a genuine non-Horn clause see the axioms for inequality in section 8.1.

Chapter 6

16.

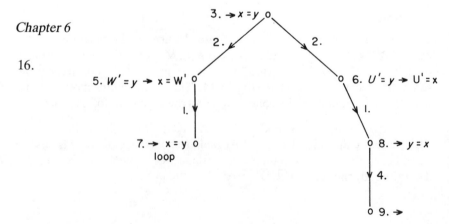

17.

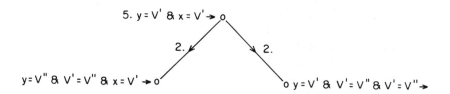

18. The complete tree is big and repetitious. I don't want to have to draw it. If you don't either then do exercises 66, 64 and 65 below, and get a computer to draw it for you.

19. The node in the bottom right hand corner, with score 9, should be developed next.

Chapter 7

20. We need the axioms:

13. $X \epsilon G$ & exponent$(g,2) \rightarrow i(X)$
14. $\rightarrow$ el$(G)\epsilon G$
15. $\rightarrow$ el'$(G)\epsilon G$
16. $X \epsilon G$ & $Y \epsilon G \rightarrow X o Y \epsilon G$
17. $X \epsilon G$ & exponent$(g,2)(G) \rightarrow X o X = e(G)$

The proof continues as follows:

12. $[i(el(g)) o i(el'(g))] o [el(g) o el'(g)] = e(g)$
        paramodulate with 13.
18. $[el(g) o i(el'(g))] o [el(g) o el'(g)] = e(g)$
        & el$(g)\epsilon g$ & exponent$(g,2) \rightarrow$
        paramodulate with 13.
19. $[el(g) o el'(g)] o [el(g) o el'(g)] = e(g)$
        & el$(g)\epsilon g$ & exponent$(g,2)$
        & el'$(g)\epsilon g$ & exponent$(g,2) \rightarrow$
            resolution with 17.

20. el(g) o el'(g) ε g & exponent(g,2)
      & el(g)εg & exponent(g,2)
      & el'(g)εg & exponent(g,2) →
  resolve with 6.

21. el(g) o el'(g) ε g
      & el(g)εg
      & el'(g)εg →
          resolve with 16.

22. el(g)εg & el'(g)εg
      & el(g)εg
      & el'(g)εg →
          resolve with 14.

23. el'(g)εg & el'(g)εg →
          resolve with 15.

24. →

Chapter 8

21.

$$5.X < 11 \rightarrow 7.X < 16$$
      close
$$\forall X\ 5.X < 11 \rightarrow 7.X < 16$$
      negate
$$\sim\forall X\ 5.X < 11 \rightarrow 7.X < 16$$
      Skolemize
$$5.x < 11\ \&\ \sim 7.x < 16$$
      This is in disjunctive normal form

22.
$$5.x + e < 11\ \&\ 16 \leqslant\ 7.x$$

23.
$$x < 11/5 - e\ \&\ 16/7 \leqslant x$$

24.
$$16/7 \leqslant x \leqslant 11/5 - e$$

However, 16/7 is greater than 11/5, (since $16.5 = 80 > 77 = 11.7$) so [16/7, 11/5) is not a possible type for x.

25. Putting the clauses in disjunctive normal gives:

$$[\sim 2 + e \leqslant a\ \&\ \sim b + e \leqslant 2\ \&\ \sim 0 \leqslant 5]$$
      v
$$[\sim 2 + e \leqslant a\ \&\ \sim b + e \leqslant 2\ \&\ \sim a \leqslant 5]$$

Eliminating $\sim$ and the resulting $<$ signs gives:

$$[a{\leqslant}2 \ \& \ 2{\leqslant}b \ \& \ 5{-}\epsilon{\leqslant}0]$$
$$\text{v}$$
$$[a{\leqslant}2 \ \& \ 2{\leqslant}b \ \& \ 5{-}\epsilon{\leqslant}a]$$

The first of these disjuncts is contradictory, since $5{+}\epsilon{\leqslant}0$ evaluates to f. 'Solving' for a in the second disjunct gives 'solution'

$$5{+}\epsilon \ \leqslant \ a \ \leqslant \ 2$$

which assigns the impossible 'type' (5,2] to a.

Chapter 9

26. $\sim\sim A \Rightarrow A$ applies to

$\sim p \ \text{v} \sim\sim q \ \text{v} \ r$

and gives

$\sim p \ \text{v} \ q \ \text{v} \ r$

The value of:

exp	is	$\sim p \ \text{v} \sim\sim q \ \text{v} \ r$
sub	is	$\sim\sim q$
lhs	is	$\sim\sim A$
rhs	is	A
φ	is	$\{q/A\}$

27. Here are 3 arguments:

(a)Each time the rule is applied the size of the expression tree is reduced by 2 plus the size of the expression substituted for X.

(b)Each time the rule is applied the number of occurrences of . decreases by 1.

(c)Each time the rule is applied a $+$ is moved below a ..

Each of these define a numerical measure, which is reduced by each application, and cannot be negative.

28. Y will unify with X.0. Hence, both rules will apply to any instance of (X.0).1, e.g. (3.0).1.

Chapter 10

29. The variable free instances of $seg(d,a){=}seg(d,X) \to$ are:

1. $seg(d,a){=}seg(d,a) \to$
2. $seg(d,a){=}seg(d,b) \to$

3. seg(d,a)=seg(d,c) →
4. seg(d,a)=seg(d,d) →

Only 1. and 3. pass the truth test.

30. The interpretation, arith6, in which . and $/$ have their normal meanings but a is assigned the value 6, has the desired effect.

31. Any interpretation in which a and b are assigned two un= values, e.g. the symmetric group of two elements in which a is assigned 0 and b is assigned 1. a=b is false in this model so a=b → fails the truth test.

32. Consider the interpretation num3. → is-power-3(a) is true in (and rejected by) num3, and this truth is inherited by → is-prod-primes(a). But a false clause (the empty clause) is then derived from it by ancestor resolution with a false ancestor, is-prod-primes(a) →. Similar remarks hold for the other branch and for the other two interpretations, num2 and num9.

Chapter 11

33.

<nil> is cons(nil,nil)

<nil,nil,nil> is cons(nil,cons(nil,cons(nil,nil)))

34.

equal((x+y)+s(z), x+(y+s(z))) = *tt* →
            This is rewritten by Symbolic Evaluation as follows:
   equal(s((x+y)+z), x+(y+s(z))) = *tt* →
   equal(s((x+y)+z), x+s(y+z)) = *tt* →
   equal(s((x+y)+z), s(x+(y+z))) = *tt* →
            by repeated use of the rewrite rule X+s(Y) => s(X+Y).
            This is then Fertilized with the induction hypothesis to give:
   equal(s(x+(y+z)), s(x+(y+z))) = *tt* →
            which evaluates to:
   tt = *tt* →

35.

(X+s(Y))+Z = X+(s(Y)+Z)

would be put in Boyer/Moore notation, negated and Skolemized to the goal clause:

equal((x+s(y))+z, x+(s(y)+z)) = *tt* →

Symbolic Evaluation would fail on this, but would find the unflawed, Generalization candidate, s(y). Generalization would convert the problem to:

$$\text{equal}((x+w)+z, x+(w+z)) = tt \rightarrow$$

which would be solved as before.

36.

$$(2^{x^2})x^3 = 2.$$

Attracting the two occurrences of x with $(U^V)^W => U^{V.W}$ gives:

$$2^{x^2} . x^3 = 2.$$

Collecting the two occurrences with $U^V.U^W => U^{V+W}$ gives:

$$2^{x^2+3} = 2.$$

And Isolating plus arithmetic evaluation gives:

$$x^5 = 1.$$
$$x = 1.$$

37. <2,2>

38. In the selective context there will a single occurrence of the unknown on the left hand side. Let the depth of this occurrence be d. Each application of Isolation produces a disjunction of equations in which d is reduced by 1. Since d can never be negative Isolation must terminate.

39.

$$\text{isolate(Posn,Old1)} = \text{Ans \& isolate(Posn,Old2)} \equiv \text{Ans}$$
$$\rightarrow \text{isolate(Posn,Old1 v Old2)} \equiv \text{Ans1 v Ans2}$$

40. All occurrences of $\equiv$ are object-level except those between an *occ* function and a number, e.g.

1) $\text{occ}(X,A=B)\underset{\text{o}}{=}1 \,\&\, \text{expr-at(List,A)} \underset{\text{m}}{\equiv} X \,\&$

$\text{isolate(List, A=B)} \underset{\text{o}}{\equiv} \text{Ans} \rightarrow \text{solve}(A=B,X,\underset{\text{o}}{\text{Ans}})$

All occurrences of $\equiv$ are meta-level

Chapter 13

41.

generalization(C, GC)
example(C, Ex)
---
example(GC, Ex)

Chapter 14

42.

distance = 400.miles

43.

     mass(stone, mass0)
     measure(5, oz, mass0)
     accel(stone, acc0, 270)
     measure(32, ft/sec^2, acc0)

44.

o   *accel(man, quickly, down)* $\rightarrow$

      (i)

o   accel(Part2, quickly, Dir2) &
     pulley-sys(Pull, Str, man, Part2) &
     extensibility(Str, 0) &
     end(Str, End1, left) &
     incline(Str, End1, down) $\rightarrow$

      (ii)

o   accel(Part2, quickly, Dir2) &
     pulley-sys(Pull, Str, man, Part2) &
     pulley-sys(Pull, Str, man, Part2) &
     end(Str, End1, left) &
     incline(Str, End1, down) $\rightarrow$

      (iii)

o   *accel(bucket, quickly, Dir2)* &
     end(rope, End1, left) &
     incline(rope, End1, down) $\rightarrow$
      (iv)

o   *end(rope, End1, left)* &
     incline(rope, End1, down) $\rightarrow$

      (v)

o   *incline(rope, knot, down)* $\rightarrow$

      (vi)

o $\rightarrow$

Chapter 15

45.

$\forall X (A) \to B$	=>	$\exists X (A \to B)$
$\exists X (A) \to B$	=>	$\forall X (A \to B)$
$A \to \forall X (B)$	=>	$\forall X (A \to B)$
$A \to \exists X (B)$	=>	$\exists X (A \to B)$

It is a little strange that quantifiers in front of the A change type, but those in front of B do not. This also prevents us from formulating rules for $\longleftrightarrow$ solely in terms of $\longleftrightarrow$.

46. Consider, for instance, the rule

$$\forall X (A) \& B => \forall X (A \& B)$$

Assume that B does not contain X (otherwise substitute a new variable for X in $\forall X$ A). Now

$\forall X (A) \& B$ is true in an interpretation I
   iff
Both $\forall X$ A and B are true in I.
   iff
For all c in the universe of I, A(c) is true in I and B is true in I.
   iff
For all c in the universe of I, A(c) & B is true in I.
   iff
$\forall X$ (A&B) is true in I.

The other examples are very similar.

47.

$$|X| < \text{delta}(M) \to 1/X > M$$

48.

new-skolem(A$\longleftrightarrow$B, Vars, Par)
 = new-skolem(A$\to$B & B$\to$A, Vars, Par)
 = new-skolem(A$\to$B, Vars, Par) &
   new-skolem(B$\to$A, Vars, Par)
 = [new-skolem(A, Vars, opposite(Par))
   $\to$ new-skolem(B, Vars, Par)] &
  [new-skolem(B, Vars, opposite(Par))
   $\to$ new-skolem(A, Vars, Par)]

However the difference in the parities now prevents us reconstructing this as an application of *new-skolem* to A⟷B.

49.

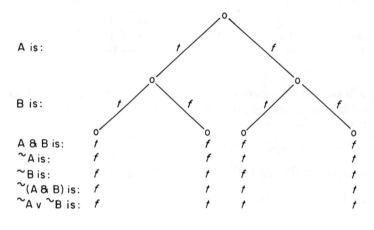

A & B is:	*t*	*f*	*f*	*f*
~A is:	*f*	*f*	*t*	*t*
~B is:	*f*	*t*	*f*	*t*
~(A & B) is:	*f*	*t*	*t*	*t*
~A v ~B is:	*f*	*t*	*t*	*t*

50.

(~p v q v p) & (~r v p)

51.

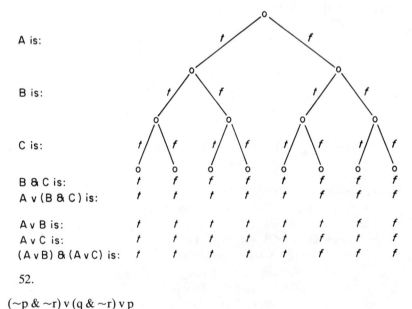

B & C is:	*t*	*f*	*f*	*f*	*t*	*f*	*f*	*f*
A v (B & C) is:	*t*	*t*	*t*	*t*	*t*	*f*	*f*	*f*
A v B is:	*t*	*t*	*t*	*t*	*t*	*t*	*f*	*f*
A v C is:	*t*	*t*	*t*	*t*	*t*	*f*	*t*	*f*
(A v B) & (A v C) is:	*t*	*t*	*t*	*t*	*t*	*f*	*f*	*f*

52.

(~p & ~r) v (q & ~r) v p

Chapter 16

53.

1. unsats(resolvant(N<S)) → unsats(S)
2. feS' → unsats(S')
3. → f ε resolvants(n,s)
4. unsats(s) →

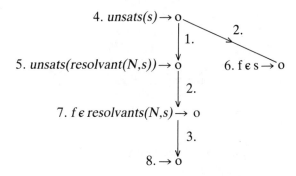

4. *unsats(s)* → o
    1.
    2.

5. *unsats(resolvant(N,s))* → o   6. *f ε s* → o
    2.

7. *f ε resolvants(N,s)* → o
    3.

8. → o

54.

Soundness: ~unsats(S) → ~unsats(resolvants(1,S))

Completeness: unsats(S) → ∃N f ε resolvants(N,S)

From which we can derive:
unsats(S) ⟷ ∃N f ε resolvants(N,S)

Chapter 17

55.

   (a) {1+2/X, U+2/Y, V/Z, 3+2/W}
   (b) {1+2/X, U+2/Y, 3+2/W}
   (c) {1+2/X, U+V/Y, 3+2/W}

56.

a) succeeds {2/X}
b) fails
c) succeeds {a/X, g(b)/Y}
d) fails

57.

a) succeeds {a/X, b/Y}
b) fails

c) succeeds {f(a)/X, a/Y}

d) fails

e) fails

58.

r(a+(Y+b), a+U, V+b) →
          paramodulate (U'+V')+W'=U'+(V'+W') into Y+b
r(a+(U'+(V'+b)), a+U, V+b) →
          paramodulate (U"+V")+W"=U"+(V"+W")
                into a+(U'+(V'+b))
r((a+U')+(V'+b), a+U, V+b) →
          resolve with r(X+Z, X, Z)
→

59.

assoc-unify X+Z and a+(Y+b) given {}
          choose {a/X}
assoc-unify a+Z and a+(Y+b) given {a/X}
          choose {Y+b/Z}
assoc-unify a+(Y+b) and a+(Y+b) given {a/X, Y+b/Z}

assoc-unify X+Z and a+(Y+b) given {}
          choose {a+U/X} and normalize
assoc-unify a+(U+Z) and a+(Y+b) given {a+U/X}
          choose {Y/U}
assoc-unify a+(Y+Z) and a+(Y+b) given {a+Y/X, Y/U}
          choose {b/Z}
assoc-unify a+(Y+b) and a+(Y+b) given {a+Y/X, Y/U, b/Z}

60.

p(g(Y, a+Y), a+a) →
          paramodulate (U+V)+W=U+(V+W) into a+Y
p(g(V+W, (a+V)+W), a+a) →
          resolve with → p(g(X, X+a), X)
          →

Chapter 18

61. The student is taking the positive difference between numbers in the same column without regard for the order of subtraction. This behaviour

can be produced from the clauses of figure 18-3 by deleting clause 4) and the proposition subtrahend(Sum1) <(minuend(Sum1) from clause 5).

62.

    append(<one,thing>, Back, <one,thing,and,another>) →
        2   {one/Car, <thing>/Cdr, Back/List,
                <thing,and,another>/Ans}
    append(<thing>, Back, <thing,and,another>) →
        2   {thing/Car, <>/Cdr, Back/List,
                <and,another>/Ans}
    append(<>, Back, <and,another>) →
        1   {Back/List, <and,another>/Back}

Appendix I
63.

    :– equal(y,x).
                (iv)
    :– equal(y,V), equal(x,V).
                (iii)
    :– equal(x,y).
                (iv)
    :– equal(x,V'), equal(y,V').
                (iii)
    :– equal(y,x).
                (iv)
    :– equal(y,V"), equal(x,V").

64. Add the new procedure *input(C)*, which is true iff C is an input clause, and modify *successor* as below.

    input(reflexive).
    input(twisted).
    input(hypothesis).

    successor(Current,Clause) :-
            factor(Current,Clause).

    successor(Current,Clause) :-
            input(Parent)
            is-clause(Parent,—,—),
            (resolve(Current,Parent,Clause) ;
                resolve(Parent,Current,Clause)).

65. Well it worked when I did it. Are you sure you typed it in correctly?

Does your version of PROLOG provide slightly different syntax or evaluable predicates? No? Well it must be software rot!

66. See the definition of *message* and its insertion into *successor* in 'The Complete Program' above in Appendix I.

67. See the solution for exercise 67 above.

68. See the definition of *in* and its insertion into *add-to-agenda* in 'The Complete Program' above in Appendix I. Note that *in* (*clause*) assumes that clause and all clauses in the data base are variable free. Otherwise, the loop check would prevent two *unifiable* clauses being in the data base.

69.

```
/* Clauses */

clause(right, [not-div(X*Z,Y)], [not-div(X,Y)]).
clause(left, [not-div(Z*X,Y)], [not-div(X,Y)]).
clause(thirty, [equal(30,2*3*5)], []).
clause(hypothesis, [not-div(5,a)], []).
clause(hypothesis, [not-div(5,a)], []).
clause(conclusion, [], [not-div(30,a)]).

/* Models */


/* arith2 */

interpret(arith2, 2, a).
interpret(arith2, N, N) :- integer(N).
interpret(arith2, false, not-div(X,Y)) :-
        0 is Y mod X, !.
interpret(arith2, true, not-div(X,Y)).
interpret(arith2, true, equal(X,Y)) :-
        X == Y.
interpret(arith2, false, equal(X,Y)) :-
        X\ == Y.
interpret(arith2, Z, X*Y) :- Z is X*Y.
```

```
/* arith3 */

interpret(arith3, 3, a).
interpret(arith3, N, N) :- integer(N).

interpret(arith3, false, not-div(X,Y)) :-
        0 is Y mod X, !.
interpret(arith3, true, not-div(X,Y)).i
interpret(arith3, true, equal(X,Y)) :-
        X == Y.
interpret(arith3, false, equal(X,Y)) :-
        X \== Y.
interpret(arith3, Z, X*Y) :- Z is X*Y.
```

Bibliography

[Aubin 75]
Aubin, R. Some generalization heuristics in proofs by induction. In Huet, G. and Kahn, G., editor, *Actes du Colloque Construction: Amelioration et verification de Programmes.*, 1975.

[Bartlett 67]
Bartlett. *Remembering*. Cambridge Univ. Press, 1967.

[Bledsoe & Hines 80]
Bledsoe, W.W. and Hines, L.M. *Variable elimination and chaining in a resolution-based prover for inequalities*. Memo ATP-56a, Math. Dept., U. of Texas, April, 1980.

[Bledsoe 74]
Bledsoe, W.W. *The Sup-Inf method in Presberger Arithmetic*. Memo ATP-18, Math. Dept., U. of Texas, 1974.

[Bledsoe 77]
Bledsoe, W.W. Non-Resolution theorem-proving. *Artificial Intelligence* 9(1):1-35, August, 1977.

[Bobrow 64]
Bobrow, D. Natural Language input for a computer problem solving system. In Minsky, M., editor, *Semantic information processing*, pages pp146-226. MIT Press, 1964.

[Borning and Bundy 81]
Borning, A and Bundy, A. Using matching in algebraic equation solving. In Schank, R., editor, *IJCAI7*, pages 466-471. International Joint Conference on Artificial Intelligence, 1981. Also available from Edinburgh as DAI Research Paper No. 158.

[Boyer & Moore 79]
Boyer, R.S. and Moore, J.S. *ACM monograph series. : A Computational Logic*. Academic Press, 1979.

[Boyer and Moore 73]
Boyer, R.S. and Moore J.S. Proving theorems about LISP functions. In Nilsson, N., editor, *procs. of IJCAI3*, pages 486-493. Stanford, August, 1973. Also available from Edinburgh as DCL memo no. 60.

[Brown and Burton 78]
Brown, J.S. and Burton, R. Buggy. *Cognitive Science* 2:155-192, 1978.

[Bundy and Silver 81]
Bundy, A. and Silver, B. Homogenization: Preparing Equations for Change of Unknown. In Schank, R., editor, *IJCAI7*. International Joint Conference on Artificial Intelligence, 1981. Longer version available from Edinburgh as DAI Research Paper No. 159.

312

[Bundy and Sterling 81]
Bundy, A. and Sterling L.S. *Meta-level Inference in Algebra*. Research Paper 164, Dept. of Artificial Intelligence, Edinburgh, September, 1981. Presented at the workshop on logic programming for intelligent systems, Los Angeles, 1981.

[Bundy and Welham 81]
Bundy, A. and Welham, B. Using meta-level inference for selective application of multiple rewrite rules in algebraic manipulation. *Artificial Intelligence* 16(2), 1981.

[Bundy et al 79a]
Bundy, A., Byrd, L., Luger, G., Mellish, C., Milne, R. and Palmer, M. Solving Mechanics Problems Using Meta-Level Inference. *In Procs of IJCAI-79*, pages 1017-1027, 1979. Also available from Edinburgh as DAI Research Paper No. 112.

[Bundy et al 79b]
Bundy, A., Byrd, L., Luger, G., Mellish, C., Milne, R. and Palmer, M. *Mecho: A program to solve Mechanics problems*. Working Paper 50, Dept. of Artificial Intelligence, Edinburgh, 1979.

[Chang and Lee 73]
Chang C-L. and Lee R. C-T. *Symbolic logic and mechanical theorem proving*. Academic Press, 1973.

[Church 40]
Church, A. A formulation of the simple theory of types. *Symbolic Logic* 5(1):56-68, 1940.

[Clocksin and Mellish 81]
Clocksin, W.F. and Mellish, C.S. *Programming in Prolog*. Springer Verlag, 1981.

[Coelho et al 80]
Coelho, H., Cotta, J.C. and Pereira, L.M. *How to solve it with PROLOG*. Technical Report, Laboratorio Nacional de Engenharia Civil, Lisbon 1980.

[Cooper 72]
Cooper, D.C. Theorem proving in arithmetic without multiplication. In Meltzer, B. and Michie, D., editor, *Mach. Intell. 7*, pages 91-99. Elsevier, New York, 1972.

[Cunningham 78]
Cunningham, J. An implementation of MERLIN, an analogical reasoning system. Master's thesis, University of Essex, 1978.

[Funt 73]
Funt, B. V. A procedural approach to constructions in Euclidean geometry. Master's thesis, University of British Columbia, 1973.

[Gelernter et al 63]
Gelernter, H. et al. Empirical explorations of the Geometry theorem-proving machine. In Feigenbaum and Feldman, editor, *Computers and Thought*, pages 153-63. McGraw Hill, 1963.

[Gelernter 63]
Gelernter, H. Realization of a Geometry theorem-proving. In Feigenbaum and
Feldman, editor, *Computers and Thought*, pages 134-52. McGraw Hill, 1963.

[Gilmore 60]
Gilmore, P.C. A proof method for quantificational theory. *IBM J Res. Dev.*
4:28-35, 1960.

[Gilmore 70]
Gilmore, P.C. An examination of the Geometry theorem-proving machine.
Artificial Intelligence 1:171-87, 1970.

[Goldberg & Suppes 74]
Goldberg, A. and Suppes, P. *Computer-assisted instruction in elementary logic
at the university level.* Technical Report 239, Institute of mathematical studies in
the social sciences, Stanford University, 1974.

[Goldberg 73]
Goldberg, A. *Computer assisted instruction: The application of theorem proving
to adaptive response analysis.* PhD thesis, Stanford, May, 1973. Also published
as IMSSS Stanford Technical Report 203.

[Herbrand 30]
Herbrand, J. Researches in the theory of demonstration. In van Heijenoort, J,
editor, *From Frege to Goedel: a source book in Mathematical Logic, 1879-1931,*
pages 525-81. Harvard Univ. Press, Cambridge, Mass, 1930.

[Hill 74]
Hill, R. *Lush-Resolution and its completeness.* DCL Memo 78, Dept. of
Artificial Intelligence, Edinburgh, August, 1974.

[Huet & Oppen 80]
Huet, G. and Oppen, D.C. Equations and rewrite rules: a survey. In Book, R.,
editor, *Formal languages: perspectives and open problems,.* Academic Press,
1980. Presented at the conference on formal language theory, Santa Barbara,
1979. Available from SRI International as technical report CSL-111.

[Huet 74]
Huet, G.P. *A unification algorithm for typed lambda-calculus.* note de travail
A 055, Institut de Recherche d'Informatique et d'Automatique, March, 1974.

[Huet 77]
Huet, G. *Confluent reductions: Abstract properties and applications to term
rewriting systems.* Rapport de Recherche 250, Laboratoire de Recherche en
Informatique et Automatique, IRIA, France, August, 1977.

[Knuth and Bendix 70]
Knuth, D.E. and Bendix, P.B. Simple word problems in universal algebra. In
Leech, editor, *Computational problems in abstract algebra,* pages 263-297.
Pergamon Press, 1970.

[Kowalski and Kuehner 71]
Kowalski, R.A. and Kuehner, D. Linear Resolution with selection function.
Artificial Intelligence 2:227-60, 1971.

[Lakatos 76]
Lakatos, I. *Proofs and refutations: The logic of Mathematical discovery.* Cambridge University Press, 1976.

[Lenat 77a]
Lenat, D.B. Automated theory formation in Mathematics. In Reddy, R., editor, *Procs of IJCAI-77,* pages 833-842., August, 1977.

[Lenat 77b]
Lenat, D.B. The ubiquity of discovery. In Reddy, R., editor, *Procs of IJCAI-77,* pages 1093-1105., August, 1977.

[Lenat 82]
Lenat D.B. AM: An Artificial Intelligence approach to discovery in Mathematics as Heuristic Search. *In Knowledge-based systems in artificial intelligence..* McGraw Hill, 1982. Also available from Stanford as TechReport AIM 286.

[Loveland 78]
Loveland, D.W. *Fundamental studies in Computer Science.* Volume 6: *Automated theorem proving: A logical basis.* North Holland, 1978.

[Marples 74]
Marples, D. *Argument and technique in the solution of problems in Mechanics and Electricity.* CUED/C-Educ/TRI, Dept. of Engineering, Cambridge, England, 1974.

[Mathlab 77]
Mathlab Group. *MACSYMA Reference Manual.* Technical Report, MIT, 1977.

[McCarthy et al 62]
McCarthy, J., Abrahams, P.W., Edwards, J.E., Hart, T.P. and Levin, M.J. *LISP 1.5 Programmers Manual.* The MIT Press, 1962.

[Mellish 80]
Mellish, C.S. *Coping with uncertainty: Noun phrase interpretation and early semantic analysis.* PhD thesis, Dept of Artificial Intelligence, University of Edinburgh, 1980.

[Mendelson 64]
Mendelson, E. *Introduction to Mathematical Logic.* van Nostrand Reinhold Co., 1964.

[Moore and Newell 73]
Moore, J. and Newell, A. How can Merlin understand. In Gregg, L., editor, *Knowledge and Cognition,* pages 201-252. Lawrence Erlbaum Associates, 1973.

[Moore 74]
Moore, J. *Computational Logic: Structure sharing and proof of program properties, part II.* PhD thesis, Univ. of Edinburgh, 1974. Available from Edinburgh as DCL memo no. 68 and from Xerox PARC, Palo Alto as CSL 75-2.

[Moses 67]
Moses, J. *Symbolic integration.* PhD thesis, MIT, December, 1967. available as MAC-TR-47.

[Nilsson 80]
   Nilsson, N.J. *Principles of Artificial Intelligence*. Tioga Pub. Co., Palo Alto,
   California, 1980.

[O'Shea and Young 78]
   O'Shea, T. and Young, R. *A production rule account of errors in children's
   subtraction*. Working Paper 42, Dept. of Artificial Intelligence, Edinburgh,
   October, 1978.

[Plotkin 72]
   Plotkin, G. Building-in equational theories. In Michie, D and Meltzer, B, editor,
   Machine Intelligence 7. Edinburgh University Press, 1972.

[Polya 45]
   Polya, G. *How to solve it*. Princeton University Press, 1945.

[Polya 65]
   Polya, G. *Mathematical discovery*. John Wiley & Sons, Inc, 1965. Two volumes.

[Raulefs et al 78]
   Raulefs, P.,Siekmann, J.,Szabo, P. and Unvericht, E. A short survey on the state
   of the art in matching and unification problems. *AISB Quarterly* issue
   32:pp17-21, December, 1978.

[Richter 74]
   Richter, M. *A note on paramodulation and the functional reflexive axioms*.
   Technical Report, Univ. of Texas at Austin, 1974.

[Robinson and Wos 69]
   Robinson, G. and Wos, L. Paramodulation and Theorem-proving in first-order
   theories with equality. In Michie, D., editor, *Machine Intelligence 4*, pages
   103-33. Edinburgh University Press, 1969.

[Robinson 65]
   Robinson, J.A. A machine oriented logic based on the Resolution principle. *J
   Assoc. Comput. Mach.* 12:23-41, 1965.

[Schubert 76]
   Schubert, L.K. Extending the expressive power of Semantic Networks. *Artificial
   Intelligence* 7:pp 89-124, 1976.

[Shostak 77]
   Shostak, R.E. On the SUP-INF method for proving Presburger formulae. *JACM*
   24(4):pp529-543, October, 1977.

[Shostak 79]
   Shostak, R.E. A practical decision procedure for arithmetic with function
   symbols. *JACM* 26(2):pp351-360, April, 1979.

[Slagle 63]
   Slagle, J.R. A heuristic program that solves symbolic integration problems in
   freshman calculus. In Feigenbaum, E.A. and Feldman, J., editor, *Computers
   and Thought*, pages 191-203. McGraw Hill, 1963.

[Sterling et al 82]
Sterling, L., Bundy, A., Byrd, L., O'Keefe, R., and Silver, B. Solving Symbolic Equations with PRESS. In Calmet, J. (editor), *Computer Algebra, Lecture Notes in Computer Science No. 144.*, pages 109-116. Springer Verlag, 1982. Longer version available from Edinburgh as Research Paper 171.

[Waerden 71]
Van der Waerden, B. L. How the proof of Baudet's conjecture was found. In Mirsky, L. (editor), *Papers presented to Richard Rado on the occasion of his sixty-fifth birthday,* pages 252-260. Academic Press, London and New York, 1971.

[Winograd 72]
Winograd, T. *Understanding Natural Language.* Edinburgh University Press, 1972.

[Wos et al 65]
Wos, L., Robinson, G. and Carson, D.F. The automatic generation of proofs in the language of Mathematics. In *Proceedings of IFIP Congress 65*, Volume 2, pages 325-326. Barton Books, Washington D.C., 1965.

[Wos 82]
Wos, L. Solving open questions with an automated theorm-proving program. In Loveland, D., editor, *Proceedings of CADE,* pages 1-31. Springer, 1982.

Index

Terms being defined are in italics in the body of the text and their page references below are in bold face type.